Progress in Mathematics
Vol. 48

Edited by
J. Coates and
S. Helgason

Birkhäuser Verlag
Boston · Basel · Stuttgart

A. Fröhlich

Classgroups and Hermitian Modules

1984

Birkhäuser
Boston · Basel · Stuttgart

Author:

A. Fröhlich
Mathematics Department and Mathematics Department
Imperial College, London, England
Robinson College, Cambridge, England

Library of Congress Cataloging in Publication Data

Fröhlich, A. (Albrecht), 1916–
Classgroups and Hermitian modules.

(Progress in mathematics ; vol. 48)
Bibliography: p.
Includes index.
1. Class groups (Mathematics) 2. Modules (Algebra)
I. Title. II. Series: Progress in mathematics (Boston, Mass.) ; vol. 48.
QA247.F7583 1984 512'.74 84-11109
ISBN-13: 978-1-4684-6742-0

CIP-Kurztitelaufnahme der Deutschen Bibliothek

Fröhlich, Albrecht:
Classgroups and hermitian modules / A. Fröhlich.
– Boston ; Basel ; Stuttgart : Birkhäuser, 1984.
(Progress in mathematics ; Vol. 48)
ISBN-13: 978-1-4684-6742-0

NE: GT

Softcover reprint of the hardcover 1st edition 1984

ISBN-13: 978-1-4684-6742-0 e-ISBN-13: 978-1-4684-6740-6
DOI: 10.1007/978-1-4684-6740-6

9 8 7 6 5 4 3 2 1

To

Ruth, Sorrel and Shaun

An earlier version of these notes has been circulated and quoted under the title "Classgroups, in particular Hermitian Classgroups" .

PREFACE

These notes are an expanded and updated version of a course of lectures which I gave at King's College London during the summer term 1979. The main topic is the Hermitian classgroup of orders, and in particular of group rings. Most of this work is published here for the first time.

The primary motivation came from the connection with the Galois module structure of rings of algebraic integers. The principal aim was to lay the theoretical basis for attacking what may be called the "converse problem" of Galois module structure theory: to express the symplectic local and global root numbers and conductors as algebraic invariants. A previous edition of these notes was circulated privately among a few collaborators. Based on this, and following a partial solution of the problem by the author, Ph. Cassou-Noguès and M. Taylor succeeded in obtaining a complete solution. In a different direction J. Ritter published a paper, answering certain character theoretic questions raised in the earlier version. I myself disapprove of "secret circulation", but the pressure of other work led to a delay in publication; I hope this volume will make amends. One advantage of the delay is that the relevant recent work can be included. In a sense this is a companion volume to my recent Springer-Ergebnisse-Bericht, where the Hermitian theory was not dealt with.

Our approach is via "Hom-groups", analogous to that followed in recent work on locally free classgroups. In fact our notes also include the first really systematic and comprehensive account of this approach to classgroups in general. Moreover, the theory of the Hermitian classgroup has some new arithmetic features of independent interest in themselves, and one of our aims was to elaborate on these.

I want to record my thanks to all those involved in the new Mathematics Institute in Augsburg, who took over so willingly and efficiently the physical production of these notes.

TABLE OF CONTENTS

INTRODUCTION

The original motivation for the theory described in these notes stems from the study of "Hermitian modules" over integral group rings, and more generally over orders. The forms considered are more general than those on which the main interest of topologists and K-theorists had been focused, in that now no condition of non-singularity in terms of the order (rather than the algebra) is attached. The significance of such more general forms comes in the first place from algebraic number theory: the ring of integers in a normal extension is a Galois module with an invariant form, in terms of the trace. Topologists have however also had to consider such forms.

Apart from this application, our results are of independent arithmetic interest in that they generalise classical ones on quadratic or Hermitian lattices. The central theme is the "discriminant problem" which we shall discuss in some detail later in this introduction, and the central concept for its solution is the Hermitian classgroup. Here, as already in preceding work in a purely module theoretic context (cf. [F7]) we work with locally free modules rather than with projectives, and classgroups are consistently described in terms of "Hom-groups", i.e., of groups of Galois homomorphisms, also to be discussed in some further remarks later in this introduction. It will then become worthwhile, and even unavoidable, to look systematically also at those other classgroups, which are defined before Hermitian structure is introduced, from this new general point of view and within this new convenient language. The reader whose interest is restricted to these pre-Hermitian aspects should read Chapter I and the relevant parts of Chapters IV and V.

The approach to classgroups which we are developing arose out of the investigation of the Galois module structure of algebraic integer rings in tame normal extensions and its connection with the functional equation of the Artin L-function (cf. [F7]; [F12]).

Subsequently a parallel theory came into being, in the first place in the local context, in which the Hermitian module structure became the principal object of study (cf. [F8], [F10]); here the form comes from a relative trace, more conveniently expressed however as a form into the appropriate group ring. Again the crux of the theory lies in the connections with (now) local root numbers and Galois Gauss sums. Both in the global case and in the local Hermitian case the crucial link between the arithmetic constants on the one hand and the classgroup invariants on the other is formed by the generalised resolvents, and it is at this stage that the new Hom language for classgroups becomes absolutely vital.

Let then F be a field (say of characteristic zero), F_c its algebraic closure, $\Omega_F = \mathrm{Gal}(F_c/F)$ the absolute Galois group over F. Let Γ be a finite group - which turns up as relative Galois group over F - and R_Γ the additive group of virtual characters. Then the various classgroups are to be described in terms of groups such as $\mathrm{Hom}_{\Omega_F}(R_\Gamma, G)$ for varying Ω_F-modules G. Resolvents give rise to elements of such groups, or of related ones. On the other hand if e.g., F is a numberfield with ring of integers $\mathfrak{o}$ then the classgroup $\mathrm{Cl}(\mathfrak{o}\Gamma)$ of the integral group ring $\mathfrak{o}\Gamma$ appears as a quotient of $\mathrm{Hom}_{\Omega_F}(R_\Gamma, J(F_c))$, J the idele group. The class in $\mathrm{Cl}(\mathfrak{o}\Gamma)$ of the ring of integers in a tame, normal extension of F, with Galois group Γ is then described in terms of the above Hom-group via the resolvents. Going beyond group rings, if we consider orders in a semisimple algebra A over F, we have to replace R_Γ by a corresponding object, the Grothendieck group $K_{A,F}$ of (equivalence classes of) matrix representations of A over F_c, i.e., we study Hom groups $\mathrm{Hom}_{\Omega_F}(K_{A,F}, G)$. In this language the determinant of a matrix, or more generally the reduced norm of an element a of $A^* = GL_1(A)$ or of $GL_n(A)$ is replaced by a Galois homomorphism $K_{A,F} \to F_c^*$ (multiplicative group), again called the determinant and denoted by $\mathrm{Det}(a)$. It maps the representation class χ (in the group case the character χ) given by a matrix T into the determinant $\mathrm{Det}_\chi(a) = \mathrm{Det}\, T(a) \in F_c^*$.

Even apart from its suitability for the arithmetic applications, the advantages of a consistent use of the Hom language are tremendous.

Thus the behaviour of classgroups under change of algebra or order
(going up or going down) has a very natural description and becomes
much more transparent than hitherto, in terms of Hom groups. The crucial point here is that previous descriptions - whether ideal
theoretic or idele theoretic - were formulated essentially in terms
of the simple components, i.e., of the set of irreducible representations, and these are not preserved. To give but one example, let Δ
be a subgroup of the group Γ. Then extension of scalars from $o\Delta$
(o as above) to $o\Gamma$ yields a map $Cl(o\Delta) \to Cl(o\Gamma)$, which in terms of
the Hom groups is the contravariant image under the functor Hom of
the restriction $R_\Gamma \to R_\Delta$ of characters. The other way round, the map
$Cl(o\Gamma) \to Cl(o\Delta)$ given by restriction of scalars, comes in our description from the induction $R_\Delta \to R_\Gamma$ of characters. Neither
$R_\Gamma \to R_\Delta$, nor $R_\Delta \to R_\Gamma$ can be described by considering only irreducible characters, and this is the reason why the old way of looking
at the classgroup $Cl(o\Gamma)$ was useless in this context. Precisely
analogous formalisms also apply to all the other classgroups which we
shall consider. In the case of group rings, the Hom groups admit also
multiplication by appropriate character rings, and this makes the
property of classgroups to define Frobenious modules more accessible.
We shall presently also indicate the usefulness of the Hom-language in
a unified description of discriminants for Hermitian modules.

In a different direction the presentation of classgroups by
Galois homomorphisms has led to the discovery of certain natural subgroups and quotients, which have provided new insights and helped
considerably in explicit calculations. Indeed the methods of computa-
tion which arise are far-reaching generalisations of the old fibre
diagram techniques.

Next we come to the "discriminant problem". We first consider
the well known classical situation of a quadratic lattice (X,h) -
either for numberfields or for, say, their p-adic completions. Here
h is a non-singular quadratic form over the given field F and X
a lattice over its ring o of integers spanning the underlying
vector space of h . There is a classical notion of the discriminant
of (X,h), as a fractional ideal of o , i.e., a non zero element
(locally) or an idele (globally) modulo units (units ideles) of o .
This however can be strengthened appreciably by defining the discriminant modulo unit squares (unit idele squares) (cf. [F1] - and it is

this latter, stronger concept which we want to generalise - and we discuss the problem in these terms, (although in fact a further improvement is needed as we shall indicate below). We now consider a Hermitian form h over a (semisimple) algebra A with involution, together with a locally free module X over the given order $\mathcal{A}$, where X spans the A-module underlying h. To simplify matters we shall assume that X is actually free over $\mathcal{A}$ - and indeed the general definition of the discriminant reduces to this case and all the essential features already appear. As in the classical case of quadratic lattices, one then forms the discriminant matrix $(h(x_i,x_j))$ corresponding to an $\mathcal{A}$-basis $\{x_i\}$ of X. The obvious approach would now be to define the discriminant analogously to the classical one, just using the generalisation of determinants mentioned already, i.e., as the Galois homomorphism $\mathrm{Det}(h(x_i,x_j))$ with values $\mathrm{Det}_\chi(h(x_i,x_j))$. Here however a difficulty is encountered, which is absent in the classical case. To illustrate this, suppose for the moment that A is simple, χ the corresponding irreducible representation class, and so Det_χ essentially the reduced norm. The involution is then of one of three possible types: Orthogonal, unitary or symplectic. For the first two everything works fine, but in the symplectic case the values of Det_χ on symmetric elements are squares. If e.g., A is a quaternion algebra with the standard involution then the symmetric elements a are those in the centre F, and for $a \in F^*$, $\mathrm{Det}_\chi(a) =$ reduced norm $(a) = a^2$. Thus Det_χ has a canonical square root on symmetric elements in this case, namely $\sqrt{\mathrm{Det}_\chi(a)} = a$. This generalises - via the almost ancient notion of a Pfaffian Pf(S) which we view here as attached in the first place to a matrix S symmetric under a symplectic involution. In this context Pfaffians were already introduced by C.T.C. Wall in certain cases, specifically for the classification of based skew-symmetric forms over a field (cf. [Wa2] and of Hermitian forms over a quaternion algebra (cf. [Wa3]). Just as in the description of ordinary classgroups, we want however to get away from the restriction to simple or more generally indecomposable involution algebras. We are aiming at a unified definition of a discriminant, not one "by cases", i.e. one which covers indecomposable algebras with orthogonal, unitary as well as with symplectic involutions and works smoothly for semisimple algebras when all these three types may simultaneously be involved. The correct language is again that of

Galois homomorphisms and the definition is then obtained via a generalisation of a Pfaffian essentially analogous to that given for determinants. In the classical case of a quadratic lattice our discriminant then indeed reduces to the "strong" discriminant, i.e., taken modulo unit squares, or unit idele squares. In fact however our discriminant is an even stronger invariant than might appear from this analogy - there is a further refinement involved which e.g. in the classical case of a Hermitian lattice, with respect to a non-trivial involution of the field, yields a better discriminant than the obvious one. The details are too technical to be discussed here.

The results of the Hermitian theory are of some arithmetic interest. One aspect is a deviation from Hasse Principle, best described in terms of a map from the global Hermitian classgroup into what we call the adelic Hermitian classgroup, i.e., essentially a restricted product of the local groups. This has in general both a non-trivial kernel and a non-trivial cokernel. Contrary to previous belief of experts, there is thus a genuine global aspect to the theory. In the case of ordinary quadratic lattices one of the consequences is the theorem (Hecke) that the ideal class of a discriminant is a square. A generalisation of this theorem is thus one of the Corollaries of our global-local set up. Moreover although, as pointed out initially, the forms considered here are not in general non-singular over the order, as they are in the usual Hermitian K-theory, some of our results are both new and relevant in the latter context - in particular those at the level of algebras, rather than orders.

The main arithmetic motivation comes of course from the study of rings of integers and trace forms under the action of a Galois group. The applications in this direction lead to deep results, expressing a connection with the functional equation of the Artin L-functions.

In the rapid development, over the last twelve years, of the theory of global Galois module structure one has mainly considered the ring of algebraic integers in a tame, relative Galois extension with Galois group Γ, as a module over the integral group ring $\mathbb{Z}\Gamma$. Its class was shown to be determined by the values of the Artin root number, i.e. the constants in the functional equation, for symplectic characters. Conversely, however, this class of the Galois module does not in general determine the symplectic root numbers.

I guessed early on that one would have to introduce further structure, and that this would have to be the Hermitian structure given by the trace form. The theory as developped previously by K-theorists, with applications mainly on topology, was entirely unsuitable. It could only have dealt with unimodular trace forms, i.e. with non-ramified extensions. I was thus led to develop a more general Hermitian theory and to apply it in the arithmetic context outlined above. The application is based on (i) a rule I found, which expresses Pfaffians in this particular situation in terms of resolvents and (ii) the relation between my resolvents and the Galois Gauss sums, which forms also the basis of the global Galois module theory. This general formalism and some specific partial results led me to the conjecture, that the given arithmetic Hermitian structure determines the symplectic root numbers both globally and locally. This has now been proved by Ph. Cassou-Noguès and M. Taylor, using (iii) some sharp results on certains Hermitian classgroups and (iv) a theorem on Galois Gauss sums which already formed part of Taylor's proof of the main theorem on global Galois module structure. Here we shall give the details of the aspects under (i) and (iii) above and quote the relevant theorems for (ii) and (iv) - as these really lie outside the scope of these notes.

Definitions and results will be presented on the level of generality best suited our purpose and framework, and this means not necessarily in the widest possible generality. Chapter I is "pre-Hermitian". The basic theory of the discriminant and the Hermitian classgroup is in Chapter II. In Chapter III we study the indecomposable case explicitly and in further detail. Chapter IV is concerned with change of order and in Chapter V we deal with the specific situation of group rings. There follows in Chapter VI a brief outline of the application to Hermitian Galois module structure of rings of integers.

Notation and conventions. Throughout we shall use the standard symbols $\mathbb{N}$, $\mathbb{Z}$, $\mathbb{Q}$, $\mathbb{Q}_p$, $\mathbb{R}$, $\mathbb{C}$, $\mathbb{H}$ for the set of natural numbers, the ring of integers, the fields of rationals, p-adic rationals, reals and complex numbers respectively, and for the real quaternion division algebra. All rings R have identities, preserved by homomorphisms, and acting as identities on modules; R^* is the multiplicative group of invertible elements of R, $M_n(R)$ the ring of n by n matrices over R and so $GL_n(R) = M_n(R)^*$. If Γ is a finite group, $R\Gamma$ is its group ring over R.

Throughout $\mathcal{o}$ is a Dedekind domain, F its quotient field, A a finite dimensional separable F-algebra and $\mathfrak{A}$ an order over $\mathcal{o}$ in A, i.e., spanning A - with further conditions imposed and variants of these notations - mostly self-explanatory - introduced, as required. "In principle" F is assumed to have characteristic zero, which means that all definitions and results, stated without further hypothesis are valid in this case. Frequently, but not always, they remain valid in other characteristics. But we do not want to clutter up the exposition, and we leave it to the interested reader to find the true level of generality for himself.

Part of the arithmetic theory will be formulated with stronger restrictions, in terms of the three cases - the ones of real interest to us - namely (i) $\mathcal{o} = F$ (referred to as the "field case"), (ii) F a number field (i.e., finite over $\mathbb{Q}$), $\mathcal{o}$ its ring of algebraic integers (the "global case") (iii) F a local field by which we mean, unless otherwise mentioned, a finite extension of $\mathbb{Q}_p$ for some finite prime p, and $\mathcal{o}$ its valuation ring (the "local case"). Again often definitions and results extend to arbitrary pairs $\mathcal{o}$, F, provided they are suitably reworded.

If F is a field with prime divisor p, subscript $_p$ denotes completion at p. If p actually comes from a non zero prime ideal of $\mathcal{o}$, also denoted by p, i.e., if p is a finite prime then $M_p = M \otimes_{\mathcal{o}} \mathcal{o}_p$ is the completion of an $\mathcal{o}$-lattice M. If p is an infinite prime in a numberfield F, we formally set $M_p = M \otimes_{\mathcal{o}} F_p$ for $\mathcal{o}$-lattices M.

As already used earlier, the symbol F_c is the algebraic closure of a field F and $\Omega_F = \mathrm{Gal}(F_c/F)$ the absolute Galois group over F.

Propositions and Lemmata are numbered within each chapter, giving section number and ordinal - and similarly for equations. Back references without roman chapter numerals are within the given chapter. Theorems are numbered consecutively throughout these notes.

CHAPTER I. PRELIMINARIES

Here we shall introduce definitions and results which are needed subsequently, but which in themselves do not presuppose any Hermitian structure. Much of this is in principle well known.

§1. Locally free modules and locally freely presented torsion modules

Recall some notation: $\mathfrak{o}$ is a Dedekind domain, F a field, A a separable finite dimensional F-algebra and $\mathcal{A}$ an $\mathfrak{o}$-order spanning A. We shall describe the analogue of the familiar theory for projectives and for modules with finite projective resolution over $\mathcal{A}$, with "projective" replaced by "locally free".

An $\mathcal{A}$-module X (usually viewed as right $\mathcal{A}$-module) is <u>locally free</u> if (i) X is finitely generated, (ii) for all non-zero prime ideals $\mathfrak{p}$ of $\mathfrak{o}$, $X_{\mathfrak{p}}$ is free over $\mathcal{A}_{\mathfrak{p}}$ (iii) $X \otimes_{\mathcal{A}} A = X \otimes_{\mathfrak{o}} F$ is free over A - if $\mathfrak{o} \neq F$ this last condition can be omitted. In the present survey $K_0(\mathcal{A})$ then denotes the Grothendieck group of locally free $\mathcal{A}$-modules of fixed hand. The class in $K_0(\mathcal{A})$ of X will be denoted by [X] and similar notations will be used elsewhere. Every locally free $\mathcal{A}$-module has a rank, $r(\mathcal{A})$, and this defines a rank map $r = r_{\mathcal{A}}$. The classgroup $Cl(\mathcal{A})$ is then by definition Ker $r_{\mathcal{A}}$, i.e., we have an exact sequence

$$0 \to Cl(\mathcal{A}) \to K_0(\mathcal{A}) \xrightarrow{r_{\mathcal{A}}} \mathbb{Z} \to 0 . \qquad (1.1)$$

This is split via the map which takes $n \in \mathbb{N}$ into the class of the free module $A^n = A \times \quad \times A$. We thus get, on defining

$$(X) = [X] - [A^n], \quad \text{where} \quad n = r(A) ,$$

an "inverse" exact sequence

$$0 \to \mathbb{Z} \to K_o(A) \xrightarrow{(\ )} Cl(A) \to 0 . \tag{1.2}$$

An A-module M is a <u>locally freely presented torsion module</u> (abbreviated "l.f.p. torsion") if (i) M is o-torsion, (ii) $M \cong X/Y$ as A-module where X, Y are locally free, $X \supset Y$. Clearly then $r(X) = r(Y)$. We shall write $K_oT(A)$ for the Grothendieck group of such modules M , with respect to exact sequences. We define a homomorphism

$$\begin{cases} \delta: K_oT(A) \to K_o(A) , \\ \delta([M]) = [Y] - [X] , \quad \text{if} \quad M \cong X/Y . \end{cases} \tag{1.3}$$

By Shanuel's Lemma δ is uniquely defined. The fact that it is a homomorphism follows from

1.1 <u>Lemma:</u> <u>If</u>

$$0 \to M' \to M \to M'' \to 0$$

<u>is an exact sequence of</u> l.f.p. <u>torsion A-modules then there are locally free A-modules</u> $X \supset U \supset Y$ <u>with</u> $M \cong X/Y$, $M' \cong U/Y$, $M'' \cong X/U$.

<u>Proof:</u> Take X, Y locally free with $M \cong X/Y$, and let $U = \mathrm{Ker}\,[X \to M \to M'']$. By Shanuel's Lemma U is projective and stably locally free, hence locally free. (By the Krull Schmitt theorem, stably free is free in the complete local case!).

In the usual manner we identify $GL_n(A)$ acting on the left with the automorphism group of the right A-module A^n . We define a homomorphism $\theta_n: GL_n(A) \to K_oT(A)$. If say X is a free A-module spanning A^n , and $a \in GL_n(A)$ then $\theta_n(a) = [X/aX]$, provided that $aX \subset X$. If this last condition is not satisfied we can find $c \in \mathfrak{o} \cap F^*$ with $cX \subset X \cap aX$ and we then put $\theta_n(a) = [X/cX] - [aX/cX]$. One shows as usual that this definition is independent of the choice of either X or c within the stated conditions, and that θ_n is indeed a homomorphism. The maps θ_n are moreover consistent with the standard embeddings $GL_n(A) \to GL_{n+1}(A)$ and thus, as $K_oT(A)$ is Abelian, factorise through a homomorphism $\theta: K_1(A) \to K_oT(A)$. Let next, and just for the moment, $K_{o\,pr}(A)$ and $K_oT_{pr}(A)$ be the Grothendieck groups respectively of finitely generated projectives over A , and of A-modules of finite projective resolution which are of finite composition length over $\mathfrak{o}$. Embeddings of categories then yield homomorphisms $K_oT(A) \to K_oT_{pr}(A)$, $K_o(A) \to K_{o\,pr}(A)$. Now we have

<u>Theorem 1.</u> (i) <u>Local completions yield an isomorphism</u>

$$K_oT(A) \cong \coprod_{p} K_oT(A_p) \ .$$

<u>(product over all non zero prime ideals of $\mathfrak{o}$.)</u>

(ii) <u>The diagram</u> (<u>with</u> κ <u>coming from ring extension</u>)

$$\begin{array}{ccccccccccc}
 & & & & 0 & & 0 & & 0 & & \\
 & & & & \downarrow & & \downarrow & & \downarrow & & \\
K_1(A) & \xrightarrow{\kappa} & K_1(A) & \xrightarrow{\theta} & K_oT(A) & \xrightarrow{\delta} & K_o(A) & \xrightarrow{r} & \mathbb{Z} & \to & 0 \\
\downarrow 1 & & \downarrow 1 & & \downarrow & & \downarrow & & \downarrow & & \\
K_1(A) & \to & K_1(A) & \to & K_oT_{pr}(A) & \to & K_{o\,pr}(A) & \to & K_{o\,pr}(A) & &
\end{array}$$

commutes and has exact rows and columns, where $\mathbb{Z} \to K_o\,{}_{pr}(A)$ takes n into $[A^n]$, and where the bottom row is the known sequence for projectives.

Remark: This theorem is "in principle" known, but references to its "locally free part" are difficult to find - none are known to me. Of course the theorem can be deduced from general principles of categorical K-theory. We prefer to give a direct proof - assuming the well known results on the "projective part" of the diagram (see e.g. the discussion in [Wi], or [Ba2] (IX 6.3).)

Proof of Theorem 1. The crucial point for (i) is to show that a l.f.p. torsion A_p-module is also l.f.p. torsion as A-module. This follows by the weak approximation theorem for $\mathcal{O}$-lattices.

In (ii) the bottom row is known to be exact. The top row is clearly exact at $K_o(A)$ and at $\mathbb{Z}$, and the composite of any two maps is null. We shall establish below exactness in the local case (when indeed trivially δ is null and r an isomorphism). It follows then that in the local case $K_oT(A) \to K_o\,T_{pr}(A)$ is injective. By (i) and the corresponding result for $K_o\,T_{pr}(A)$, the map $K_o\,T(A) \to K_o\,T_{pr}(A)$ is always injective. Using now the obvious commutativity and the exactness of the bottom row, that of the top row follows. Moreover the map $\mathbb{Z} \to K_o\,{}_{pr}(A)$ is clearly injective, hence so is $K_o(A) \to K_o\,{}_{pr}(A)$.

We want to show that in the local case the map $g\colon \mathrm{Cok}\,\kappa \to K_o\,T(A)$ induced by θ is an isomorphism. For this we construct a homomorphism $f\colon K_o\,T(A) \to \mathrm{Cok}\,\kappa$, by setting $f[M] = \tilde{a}$, the class of a in $\mathrm{Cok}\,\kappa$, if $M = X/aX$, $a \in GL_q(A)$, X free of rank q. If such an f is well defined then it is the inverse of g. If indeed a module M defines a unique $\tilde{a}$ then it does follow from Lemma 1.1, that f defines a homomorphism. It thus remains to prove

1.2 Lemma: Let for $i = 1,2$, $M \cong X_i/a_iX_i$, X_i free A-modules of ranks q_i, $a_i \in GL_{q_i}(A)$. Then $\tilde{a}_1 = \tilde{a}_2$.

<u>Proof</u>: Let $h_i: X_i \to M$ be the given surjections. We get a commutative diagram with exact rows

$$\begin{array}{ccc} X_1 \oplus X_2 & \xrightarrow{\;0 \oplus h_2\;} & M \to 0 \\ \Big\downarrow t & & \Big\downarrow 1 \\ X_1 \oplus X_2 & \xrightarrow{\;h_1 \oplus h_2\;} & M \to 0 \end{array}$$

where $t(x_1,x_2) = (x_1, t'(x_1) + x_2)$, with $t': X_1 \to X_2$ being a homomorphism so that $h_2 \circ t' = - h_1$. t yields an automorphism of $X_1 \oplus X_2$, i.e.,

$$\tilde{t} = 1 \tag{1.4}$$

and thus an isomorphism of $\mathrm{Ker}(0 \oplus h_2) = X_1 \oplus a_2 X_2$ with $Y = \mathrm{Ker}(h_1 \oplus h_2)$. Therefore $b_2 = t(1 \oplus a_2)$ yields an isomorphism $X_1 \oplus X_2 \to Y$ and in view of (1.4) $\tilde{b}_2 = \tilde{a}_2$. Interchanging 1 and 2 we get an element b_1 inducing an isomorphism $X_1 \oplus X_2 \to Y$ with $\tilde{b}_1 = \tilde{a}_1$. As $b_1^{-1} b_2$ is an automorphism of $X_1 \oplus X_2$ we have $\tilde{b}_1 = \tilde{b}_2$ and therefore finally $\tilde{a}_1 = \tilde{a}_2$.

§2. <u>Determinants and the Hom language for classgroups</u>

Here we shall introduce "determinants" and the description of classgroups via Galois homomorphisms. We shall derive also a number of useful isomorphisms. Most of this is at least implicitly already in [F7] (Appendix), although our derivation of the fundamental isomorphism is here more direct than in [F7]. (See also [F12])

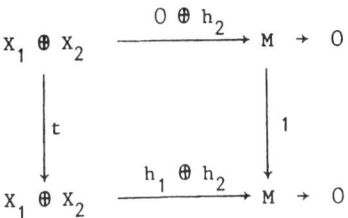

With F_c the separable closure and Ω_F the absolute Galois group $\mathrm{Gal}(F_c/F)$ of F we consider, for any separable, finite dimensional commutative F-algebra C the set $G_{C,F}$ of F-algebra homomorphisms $C \to F_c$. This is a finite Ω_F-set, where Ω_F acts by composition: $f^\omega(c) = f(c)^\omega$, for $f \in G_{C,F}$, $\omega \in \Omega_F$, $c \in C$.

2.1 Proposition: Let E be a finite extension of F in F_c and let C actually be an E-algebra. Then $G_{C,E}$ embeds (by restriction) into $G_{C,F}$ and the map

$$(f,\omega) \longmapsto f^\omega, \quad f \in G_{C,E}, \quad \omega \in \Omega_F$$

yields a bijection

$$G_{C,E} \times_{\Omega_E} \Omega_F \simeq G_{C,F},$$

where the left hand is the set of equivalence classes in $G_{C,E} \times \Omega_F$ under the relation $(f,\sigma\omega) = (f^\sigma,\omega)$ $\sigma \in \Omega_E$, $\omega \in \Omega_F$, $f \in G_{C,E}$.

Proof: Obvious.

Corollary: If X is an Ω_F-module then

$$\mathrm{Map}_{\Omega_F}(G_{C,F}, X) \simeq \mathrm{Map}_{\Omega_E}(G_{C,E}, X)$$

(group isomorphism).

Next with A as before (cf. §1) we shall write $K_{A,F} = K_0(A \otimes_F F_c)$. But we shall view this from the point of view of matrix representations i.e., of homomorphisms

$$T: A \to M_n(F_c) \tag{2.1}$$

of F-algebras. With the usual notion of equivalence:

$$T' \sim T \quad \text{if} \quad \exists\, P \in GL_n(F_c)\,, \quad T'(a) = P^{-1}T(a)P \ \forall\, a\,,$$

the equivalence classes χ_T of representations form a commutative additive semigroup, where $\chi_S + \chi_T$ corresponds to $\begin{pmatrix} S & 0 \\ 0 & T \end{pmatrix}$. Then $K_{A,F}$ is its Grothendieck group. If, say, $A = F\Gamma$ is the groupring of a finite group Γ, and F is of characteristic zero then $K_{A,F} = R_\Gamma$, the additive group of virtual characters of Γ in F_c. Let $G_{A,F}$ be the set of classes of irreducible representations. Then $K_{A,F}$ is free Abelian on the set $G_{A,F}$ and again Ω_F acts by $T,\omega \longmapsto T^\omega$ where $T^\omega(a) = T(a)^\omega$, with the action of Ω_F extended to matrices. If A is commutative this definition is essentially the same as the earlier one. More generally restriction to $C = \text{cent}\,(A)$ yields a bijection $G_{A,F} \simeq G_{C,F}$, i.e., we get

2.2. <u>Proposition</u>: <u>If</u> X <u>is an</u> Ω_F<u>-module, and</u> $C = \text{cent}(A)$, <u>then</u>

$$\text{Hom}_{\Omega_F}(K_{A,F}, X) = \text{Map}_{\Omega_F}(G_{A,F}, X) = \text{Map}_{\Omega_F}(G_{C,F}, X)\ .$$

Next we consider an embedding of fields $s\colon F \to E$, extended to an embedding $s\colon F_c \to E_c$. There is then a homomorphism $\Omega_E \to \Omega_F$, also denoted by s, with

$$s(x)^\omega = s(x^{s(\omega)})\,, \quad x \in F_c\,, \quad \omega \in \Omega_E\ .$$

We shall assume that s is injective - this is the only case we need. (Think of E as an algebraic extension, or a completion of F). Now we have

2.3 Proposition: If $g \in G_{C,F}$ then

$$s(g): C \otimes_F E \xrightarrow{sg \otimes 1} E_c \otimes_F E \xrightarrow{\text{mult}} E$$

lies in $G_{C \otimes_F E, E}$, and $g \longmapsto s(g)$ is a bijection

$$s: G_{C,F} \longrightarrow G_{C \otimes E, E} .$$

Moreover the maps

$$s^{-1}: G_{C \otimes E, E} \cong G_{C,F}$$

$$s \otimes 1: F_c \otimes_F E \longrightarrow E_c \otimes_F E \longrightarrow E_c$$

induce an isomorphism

$$s^{*}: \mathrm{Map}_{\Omega_F}(G_{C,F}, (F_c \otimes_F E)^{*}) \cong \mathrm{Map}_{\Omega_E}(G_{C \otimes E, E}, E_c^{*}) .$$

Proof: That $g \longmapsto s(g)$ maps $G_{C,F} \longrightarrow G_{C \otimes E, E}$ is clear. Restriction of maps $g': C \otimes E \to E$ to C gives the inverse map $s^{-1}: G_{C \otimes E, E} \to G_{C,F}$.

Next s^{*} is a well defined homomorphism. Moreover the Galois groups Ω act on the finite sets G via finite quotients. More precisely the homomorphisms $g: C \to F_c$ over F all factorise through a normal extension F' of F of finite degree. Viewing F_c and so F' as embedded in E_c (i.e. omitting now the symbol s), it follows that the homomorphisms $C \otimes_F E \to E_c$ will factorise through the composite field $E' = F'E$, and this field is normal over E . Its relative Galois group $\Sigma_E = \mathrm{Gal}(E'/E)$ is a subgroup of $\Sigma_F = \mathrm{Gal}(F'/F)$;

the latter is a finite group, and the action of Ω_F on $G_{C,F}$ is lifted from an action of Σ_F. All we have to prove then, in view of the first part of the proposition, is that the map

$$\hat{s}: \mathrm{Map}_{\Sigma_F}(G_{C,F}, (F' \otimes_F E)^*) \longrightarrow \mathrm{Map}_{\Sigma_E}(G_{C,F}, E'^*) ,$$

induced by $F' \otimes_F E \xrightarrow{\text{mult}} F'E = E'$, is an isomorphism. The map

$$x \otimes_F y \longmapsto [\sigma \longmapsto x^{\sigma}y],\ x \in F',\ y \in E,\ \sigma \in \Sigma_F$$

is an isomorphism $F' \otimes_F E \cong \mathrm{Map}_{\Sigma_E}(\Sigma_F, E')$ of Σ_F-algebras, where Σ_F acts by $(f\sigma)(\sigma_1) = f(\sigma\sigma_1)$. Going over to multiplicative groups, we get isomorphisms

$$(F' \otimes_F E)^* \cong \mathrm{Map}_{\Sigma_E}(\Sigma_F, E')^* \cong \mathrm{Map}_{\Sigma_E}(\Sigma_F, E'^*) .$$

Transition to maps gives us an isomorphism

$$\hat{s}_1: \mathrm{Map}_{\Sigma_F}(G_{C,F}, (F' \otimes_F E)^*) \cong \mathrm{Map}_{\Sigma_F}(G_{C,F}, \mathrm{Map}_{\Sigma_E}(\Sigma_F, E'^*)) .$$

The standard right adjoint equivalence then yields an isomorphism

$$\hat{s}_2: \mathrm{Map}_{\Sigma_F}(G_{C,F}, \mathrm{Map}_{\Sigma_E}(\Sigma_F, E'^*)) \cong \mathrm{Map}_{\Sigma_E}(G_{C,F}, E'^*) .$$

Finally one verifies that $\hat{s}$ is the compositum of $\hat{s}_1$ and $\hat{s}_2$, which gives the required result.

Remark: We get quite generally an isomorphism

$$s: K_{A;F} \cong K_{A \otimes_F E, E}$$

compatible with Galois action, which maps $G_{A,F} \cong G_{A \otimes_F E, E}$ and the latter commutes with the bijections induced by the restriction from A to $C = \text{cent}(A)$. If say $T: A \to M_n(F_c)$ is a representation then T^s is the compositum

$$A \otimes_F E \to M_n(F_c) \otimes_F E \to M_n(E_c) .$$

Next we consider functors H from finite separable extension fields L of F to Abelian groups. If Δ is a group of automorphisms of L over F we get a homomorphism

$$H(L^\Delta) \to H(L)^\Delta , \tag{2.2}$$

where X^Δ is the fixed subset in a Δ-set X. We extend H to finite direct products by $H(\Pi L_i) = \Pi H(L_i)$ and to infinite algebraic extension by direct limit. If C is a finite product of fields L of finite degree over F, we get the basic homomorphism

$$\Theta: H(C) \to \text{Map}_{\Omega_F}(G_{C,F}, H(F_c)) , \tag{2.3}$$

where $\Theta_x(g) = H(g)(x)$, for $x \in H(C)$, $g \in G_{C,F}$.

In applications this will usually appear via the identification of Proposition 2.2. as

$$\Theta: H(\mathrm{cent}(A)) \to \mathrm{Hom}_{\Omega_F}(K_{A,F}, H(F_c)) , \tag{2.3a}$$

(cf. [F7] Appendix I). A particular case of this appears also in [Wa4] (section 2.3). Now we have (cf. [F7] (A.I.3)).

2.4. Proposition: *If the maps* (2.2) *are always bijective, so are the maps* (2.3).

The result is fairly obvious when C is a field. The general case then follows by taking products.

From now on, and unless otherwise stated, let B be a commutative F-algebra, not necessarily semisimple or finite dimensional. We have

2.5 Proposition: *The maps* (2.2) *are always bijective if* H *is given by* $H(L) = (L \otimes_F B)^*$.

For, this is the case for the functor $L \longmapsto L \otimes_F B$, on algebras.

Now let $T: A \to M_n(F_c)$ be a representation. We can extend it to a homomorphism

$$\begin{aligned} M_q(A \otimes_F B) &= A \otimes_F M_q(F) \otimes_F B \to M_n(F_c) \otimes_F M_q(F) \otimes_F B \\ &= M_{nq}(F_c \otimes_F B) , \end{aligned} \tag{2.4}$$

again denoted by T . Follow this by taking determinants into the ring $F_c \otimes_F B$ and restrict to invertible elements. We then get a homomorphism of groups

$$\mathrm{Det}_\chi = \mathrm{Det}\ T: GL_q(A \otimes_F B) \to (F_c \otimes_F B)^* , \tag{2.5}$$

which indeed only depends on the class χ of T in $K_{A,F}$. Now we get

2.6. Proposition: Let $a \in GL_q(A \otimes_F B)$. The map $\chi \longmapsto \mathrm{Det}_\chi(a)$ extends to a homomorphism

$$\mathrm{Det}(a)\colon K_{A,F} \to (F_c \otimes_F B)^*$$

of groups, and the map $a \longmapsto \mathrm{Det}(a)$ is a homomorphism

$$\mathrm{Det}\colon GL_q(A \otimes_F B) \to \mathrm{Hom}_{\Omega_F}(K_{A,F}\,,\,(F_c \otimes_F B)^*)\ .$$

Moreover if $b \in GL_m(A \otimes_F B)$, a as above then

$$\mathrm{Det}\begin{pmatrix} a & 0 \\ 0 & b \end{pmatrix}(\chi) = \mathrm{Det}(a)\,(\chi)\ \mathrm{Det}(b)\,(\chi)\ .$$

Indeed the first assertion follows from the obvious relation

$$\mathrm{Det}_{\chi+\phi}(a) = \mathrm{Det}_\chi(a)\ \mathrm{Det}_\phi(a)$$

for classes χ, ϕ of representations. The second assertion is a consequence of the multiplicativity of determinants and the fact that they commute with Galois action. The last assertion is trivial.

Remark: If V is a free right A-module of rank q, then an isomorphism $A^q \cong V$ yields an isomorphism $GL_q(A) \cong \mathrm{Aut}(V)$. We can thus replace $GL_q(A)$ by $\mathrm{Aut}(V)$ in the definition of Det. The usual argument shows that the actual choice of isomorphism $A^q \cong V$ is immaterial.

By taking reduced norms in all the simple components we get a reduced norm

$$\mathrm{nrd}: GL_q(A) \to C^* , \quad C = \mathrm{cent}(A) .$$

If B is a (possibly infinite) product of fields we get, for such a product, an extension

$$\mathrm{nrd}: GL_q(A \otimes_F B) \to (C \otimes_F B)^* ,$$

and this applies also to subalgebras B of such products, e.g., B the adele ring of a number field F. Now we have (cf. [F7] (Appendix, (I.9)),

2.7 <u>Proposition</u>: <u>Let</u> $\chi \in G_{A,F}$, g_χ <u>its image in</u> $G_{C,F}$ $(C = \mathrm{cent}(A))$. <u>Then the diagram</u>

$$\begin{array}{ccc} GL_q(A \otimes_F B) & \xrightarrow{\ \mathrm{nrd}\ } & (C \otimes_F B)^* \\ & {}_{\mathrm{Det}_\chi}\searrow & \downarrow {\scriptstyle (g_\chi \otimes B)^*} \\ & & (F_c \otimes_F B)^* \end{array}$$

<u>commutes</u>.

This is in fact really the definition of the reduced norm.

We shall now use the formalism set up to derive determinantal descriptions of $K_oT(A)$ and $Cl(A)$, and of the associated maps. To avoid overloading the expositions we shall state everything in terms of number fields and their local completions. The results are in fact more general, but we shall content ourselves with indicating the appropriate general formulation, in remarks for the benefit of the reader who may be interested in this.

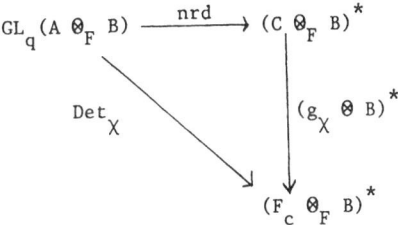

The two cases we have to consider are (i) F is a finite extension of $\mathbb{Q}_p$ for some finite p and o is its ring of integers, this always to be referred to as the <u>local case</u>, or (ii) F is a number field, i.e., a finite extension of $\mathbb{Q}$ and o its ring of algebraic integers (<u>the global case</u>). In the latter situation we shall write

$$\mathrm{Ad}(o) = \prod_p o_p \quad \text{(product over all prime divisors, finite or infinite)}$$

- recall that by convention $o_p = F_p$ for infinite p. The adele ring of F will be denoted by Ad(F). Then we put

$$JA = (A \otimes_F \mathrm{Ad}(F))^*, \quad UA = (A \otimes_o \mathrm{Ad}(o))^* .$$

Alternatively $UA = \prod_p A_p^*$, and JA is the product of the A_p^*, restricted with respect to the A_p^*. We get in particular functors $JL = \mathrm{Ad}(L)^*$ and $U(o_L) = \mathrm{Ad}(o_L)^*$ of finite algebraic extensions L of F, with o_L the integral closure of o in L. For these (suitably extended) the maps (2.2), (2.3) are always bijective. The images Det JA and Det UA, under the map Det of Proposition 2.6, are now subgroups of $\mathrm{Hom}_{\Omega_F}(K_{A,F}, J(F_c))$, and analogously in the local case. Here $J(F_c) = (\mathrm{Ad}(F) \otimes_F F_c)^*$ where Ad(F) is the adele ring, or equivalently $J(F_c)$ is the direct limit, or the union of the idele groups J(E) for finite extensions E of F. Now we get the "Hom-description", in much more detail than stated previously (cf. [F7] Appendix, I(10) and [F12])

<u>Theorem 2.</u> <u>The homomorphism</u> Det <u>sets up isomorphisms</u>

(i) $K_1A \cong \mathrm{Det}\, A^*$ <u>(in both the local and the global case)</u>,

(ii) $\mathrm{Im}(K_1 A \to K_1 A) \cong \mathrm{Det}\, A^*$ <u>(in the local case)</u>, *and hence in the local case, via the exact sequence of Theorem 1 an isomorphism*

(iii) $\quad \mathrm{Det}\, A^*/\mathrm{Det}\, A^* \cong K_oT(A)$.

Moreover in the local case

$$\mathrm{Det}\, A^* \cong \mathrm{Hom}_{\Omega_F}(K_{A,F}, F_c^*) ,$$

whence

(iv) $\quad \mathrm{Hom}_{\Omega_F}(K_{A,F}, F_c^*)/\mathrm{Det}\, A^* \cong K_oT(A)$.

In the global case

(v) $\quad \mathrm{Det}\, JA/\mathrm{Det}\, UA \cong K_oT(A)$,

and from the exact sequence of Theorem 1 *we get via* (i) *and* (v) *an exact sequence*

$$\mathrm{Det}\, A^* \to \mathrm{Det}\, JA/\mathrm{Det}\, UA \to \mathrm{Cl}(A) \to 0 ,$$

and therefore

(vi) $\quad \mathrm{Det}\, JA/\mathrm{Det}\, UA.\mathrm{Det}\, A^* \cong \mathrm{Cl}(A)$.

In fact via the embedding $\mathrm{Det}\, JA \subset \mathrm{Hom}_{\Omega_F}(K_{A,F}, J(F_c))$ *also*

(vii) $$\frac{\mathrm{Hom}_{\Omega_F}(K_{A,F}, J(F_c))}{\mathrm{Det}\, UA.\mathrm{Hom}_{\Omega_F}(K_{A,F}, F_c^*)} \cong \mathrm{Cl}(A) .$$

Remark 1. If one follows the explicit descriptions of the isomorphisms involved one finds that locally the class $[M] \in K_0T(A)$ is represented by $\mathrm{Det}(a)$, where $M \cong X/aX$, $a \in GL(A)$. Globally one gets similarly a representative $\mathrm{Det}(a)$ where now for each p, $a_p \in GL(A_p)$, with $M \cong X/Y$, X free, and $Y_p = a_pX_p$. This $\mathrm{Det}(a)$ is then also a representative of $(Y) \in Cl(A)$ (see (1.2)).

Remark 2. For a general Dedekind domain $o \neq F$ with quotient field F one gives oneself a set of prime divisors (completions) of F , including all finite ones, i.e., those corresponding to the maximal ideals of o , and possibly finitely many other prime divisors ("infinite prime divisors" - e.g., in the function field case). One can then define adele rings and idele groups and the theorem will go over - provided that the reduced norm maps K_1A isomorphically onto $\mathrm{nrd}(A)^*$.

Proof of Theorem 2. The condition stated at the end of remark 2, does indeed hold in our cases. On the other hand we get from Proposition 2.7 an isomorphism

$$\mathrm{nrd}(A^*) \cong \mathrm{Det}(A^*)$$

$$(c \longmapsto [\chi \longmapsto g_\chi(c)]) ,$$

and this yields (i). Moreover for orders over local rings, $\mathrm{nrd}(A^*) = \mathrm{nrd}\, GL_q(A)$, all q , (c.f. [SE](Prop. 8.5)) and this implies (ii). From (i), (ii) and Theorem 1. (ii) we now get (iii). The next assertion is a consequence of the fact that, in the case considered, $\mathrm{nrd}\, A^* = \mathrm{cent}\, A^*$.

Next turn to the global case. If p is a prime divisor of F we have from Propositions 2.2 and 2.3 (with $E = F_p$) the isomorphism of the top row of the diagram

$$\begin{array}{ccc} \mathrm{Hom}_{\Omega_F}(K_{A,F}, (F_c \otimes_F F_p)^*) & \cong & \mathrm{Hom}_{\Omega_{F_p}}(K_{A_p,F_p}, F^*_{p,c}) \\ & \nwarrow^{\mathrm{Det}} \quad {}^{\mathrm{Det}}\nearrow & \\ & A^*_p = (A \otimes_F F_p)^* & \end{array}$$

the other two arrows representing the two obvious determinantal maps, here not distinguished by distinct symbols. The diagram indeed commutes, and thus for finite p there is an isomorphism

$$\frac{\mathrm{Det}\ A^*_p}{\mathrm{Det}\ A^*_p} \cong K_o T(A_p) \quad ,$$

where now the symbols Det A^*_p, Det A^*_p denote subgroups of the semi local group $\mathrm{Hom}_{\Omega_F}(K_{A,F}, (F_c \otimes_F F_p)^*)$, rather than of the local group $\mathrm{Hom}_{\Omega_{F_p}}(K_{A_p,F_p}, F^*_p)$ as in (iii). Now applying Theorem 1. (i) we get (v), taking into account that the infinite component of Det JA/Det UA indeed = 1 . (vi) now follows from (i), (v) and Theorem 1. (ii).

The embedding $\mathrm{Det}\ JA \subset \mathrm{Hom}_{\Omega_F}(K_{A,F}, J(F_c))$ clearly does define a homomorphism of the left hand side of (vi) into that of (vii). To show that it is injective it suffices to show that

$$\mathrm{Det}\ JA \cap \mathrm{Hom}_{\Omega_F}(K_{A,F}, F^*_c) = \mathrm{Det}\ A^* , \tag{2.6}$$

while surjectivity follows from the equation

$$\mathrm{Det}\ JA \ . \ \mathrm{Hom}_{\Omega_F}(K_{A,F}, F^*_c) = \mathrm{Hom}_{\Omega_F}(K_{A,F}, J(F_c)) \ . \tag{2.7}$$

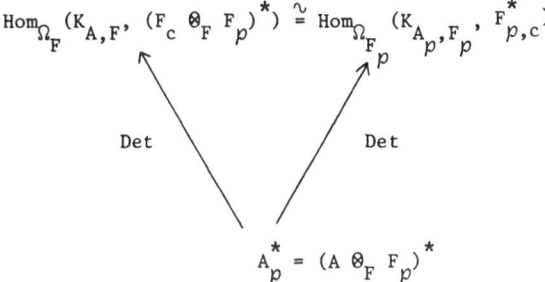

By reduction to simple components, one quickly sees that (2.7) follows from the fact that in any number field its numbers take all possible signatures at its real primes. On the other hand (2.6) is essentially equivalent with the Hasse-Schilling norm theorem, stated e.g., in [SE] (Theorem 7.6).

§3. Supplement at infinity

The description (iv) (Theorem 2) for $K_oT(A)$ also yields a possibly non-trivial group at infinite local fields, i.e., for $F = \mathbb{R}$ or $\mathbb{C}$, and thus suggests that for these fields as well we should define $K_oT(A) = K_oT(A)$ by

$$K_oT(A) = \mathrm{Hom}_{\Omega_F}(K_{A,F}, F_c^*)/\mathrm{Det}\, A^* . \tag{3.1}$$

Of course for $F = \mathbb{C}$ always $K_oT(A) = 1$. For $F = \mathbb{R}$ this is an elementary 2-group, one copy of $<\pm 1>$ appearing for each simple component of A of form $M_n(\mathbb{H})$ ($\mathbb{H}$ the real quaternion division algebra). This might suggest that $K_oT(A)$ describes "Orientations" in some space.

For the global case one can now proceed in two alternative directions. If F is a number field one can define a "torsion K_o at infinity" by

$$K_oT_\infty(A) = \prod_{p|\infty} K_oT(A_p) . \tag{3.2}$$

Using Propositions 2.2 and 2.3, as in the proof of Theorem 2, and writing J_∞ for the infinite part of the idele group, i.e. $J_\infty(F_c) = (F_c \otimes_{\mathbb{Q}} \mathbb{R})^*$, we get

$$K_oT_\infty(A) = \mathrm{Hom}_{\Omega_F}(K_{A,F}, J_\infty(F_c))/\mathrm{Det}\, J_\infty A . \tag{3.3}$$

Viewing $J_\infty(F_c)$ as subgroup of $J(F_c)$ and $\mathrm{Det}\, J_\infty A = \mathrm{Det}\, U_\infty A$ as subgroup of $\mathrm{Det}\, UA$ we derive from the isomorphism (vii) (Theorem 2) for $\mathrm{Cl}(A)$ a homomorphism

$$\delta_\infty : K_o T_\infty (A) \to \mathrm{Cl}(A) \tag{3.4}$$

$\mathrm{Ker}\, \delta_\infty$ has an interesting description. One can also consider the "combined torsion K_o", writing

$$\widetilde{K_o T(A)} = K_o T(A) \times K_o T_\infty(A) , \tag{3.5}$$

with the alternative description

$$\widetilde{K_o T(A)} = \mathrm{Hom}_{\Omega_F}(K_{A,F}, J(F_c))/\mathrm{Det}\, UA . \tag{3.6}$$

For another approach we extend the definition (3.1) to arbitrary fields F. If in particular F is a number field then indeed

$$K_o T(A) = \prod_{p|\infty} K_o T(A_p) \tag{3.7}$$

while, for finite p, $K_o T(A_p) = 1$.

CHAPTER II. INVOLUTION ALGEBRAS AND THE HERMITIAN CLASSGROUP

This chapter contains the basic theory. The main problem, namely the definition of a good discriminant for Hermitian modules, which was alluded to in the introduction, will be posed in §2 and solved in §3 - §5.

§1. Involution algebras and duality

An involution j on a ring R is an automorphism $a \mapsto a^j$ of its additive group, so that $(ab)^j = b^j\, a^j$ and so that j^2 is the identity map. Such a j yields an involution on $M_n(R)$, called its matrix extension and again denoted by j. It is defined by

$$(a^j)_{r,s} = (a_{s,r})^j \tag{1.1}$$

where $b_{r,s}$ is the r,s entry of a matrix b - in other words we transpose the matrix and let the original j act on the elements.

With A and F as in I. and with $^-$ an involution of A over F, i.e., one leaving F elementwise fixed, we call the pair $(A,{}^-)$ an <u>involution algebra</u>(over F), but by abuse of notation often denote it just by A.

From now on assume such a pair $(A,{}^-)$ given. If $T\colon A \to M_n(F_c)$ is a representation then so is $\bar{T}$, where

$$\bar{T}(a) = T(\bar{a})_t\ , \tag{1.2}$$

the subscript $_t$ denoting the transposition. This defines an involutory automorphism $\chi \longmapsto \bar{\chi}$ of $K_{A,F}$ which maps $G_{A,F}$ into itself and commutes with the action of Ω_F, and thus also defines an involutory automorphism $f \longmapsto \bar{f}$ of $\mathrm{Hom}_{\Omega_F}(K_{A,F}, .)$ where $\bar{f}(\chi) = f(\bar{\chi})$. The restriction of $^-$ to $C = \mathrm{cent}(A)$ is again an involution and the action of $^-$ respects the identification $G_{A,F} = G_{C,F}$. From the point of view of module theory, the representation $\bar{T}$ of (1.2) corresponds to the dual of the $A \otimes_F F_c$-module defined by T - say dual over F_c .

If B is a commutative F-algebra, we extend $^-$ to $A \otimes_F B$ by letting it act trivially on B , hence to $M_q(A \otimes_F B)$ and so to $GL_q(A \otimes_F B)$.

1.1 <u>Proposition</u>: <u>The map</u>

$$GL_q(A \otimes_F B) \to \mathrm{Hom}_{\Omega_F}(K_{A,F}, (F_c \otimes_F B)^*)$$

<u>preserves</u> $^-$ <u>action</u>. <u>In other words</u>

$$\mathrm{Det}_{\bar{\chi}}(a) = \mathrm{Det}_{\chi}(\bar{a}) .$$

<u>Proof</u>: From the definition (cf. (1.1)) and the fact that

$$\mathrm{Det}\ T(\bar{a})_t = \mathrm{Det}\ T(\bar{a}) .$$

As in I., $\mathcal{O}$ is a Dedekind domain with quotient field F . From now on Λ is an $\mathcal{O}$-order spanning A with $\bar{\Lambda} = \Lambda$. If X is a locally free Λ-module (say with Λ acting on the right) we write

$$\hat{X} = \mathrm{Hom}_{\Lambda}(X,\Lambda) . \qquad (1.3)$$

We consider $\hat{X}$ again as a right A-module via the involution $\bar{\ }$. In other words, if

$$< , > : \hat{X} \times X \to A$$

is the defining pairing for $\hat{X}$, then

$$<ya,x> = \bar{a}<y,x>, \quad y \in \hat{X}, \quad x \in X, \quad a \in A . \tag{1.4}$$

Next let M be a l.f.p. torsion right A-module, say given by the exact sequence

$$0 \to Y \to X \to M \to 0 , \tag{1.5}$$

with X and Y locally free, and spanning the same A-module. We then put

$$M^* = \mathrm{Hom}_A(M, A/A) , \tag{1.6}$$

again viewing M^* as a right A-module, by a rule analogous to (1.4).

1.2 Proposition: (i) With M as in (1.5), there is a "dual" exact sequence

$$0 \to \hat{X} \to \hat{Y} \to M^* \to 0 ,$$

and a natural isomorphism

$$M^* \cong \mathrm{Ext}^1_\Lambda(M,A) \ .$$

(ii) $X \longmapsto \hat{X}$, *and* $M \longmapsto M^*$ *define exact contravariant functors of the category of locally free right* Λ-*modules, resp. of the category of l.f.p. torsion right* Λ-*modules, onto itself*.

(iii) *The double dual maps*

$$X \to \hat{\hat{X}}, \quad M \to M^{**}$$

are isomorphisms.

Proof: The assertions on locally free modules are obvious. Suppose now that M is given by the exact sequence (1.5). We derive from it an exact sequence

$$\mathrm{Hom}_\Lambda(X,A) \overset{f}{\to} \mathrm{Hom}_\Lambda(Y,A) \quad \mathrm{Ext}^1_\Lambda(M,A) \to 0 \ .$$

As $X \otimes_\Lambda A = Y \otimes_\Lambda A$, f is an isomorphism, hence

$$\mathrm{Ext}^1_\Lambda(M,A) = 0 \ .$$

From the exact sequence

$$0 \to \Lambda \to A \to A/\Lambda \to 0$$

we get an exact sequence

$$\mathrm{Hom}_A(M,A) \to M^* \to \mathrm{Ext}_A^1(M,A) \to \mathrm{Ext}_A^1(M,A) \ ,$$

with the extreme terms null, i.e., we get the natural isomorphism for M^* stated under (i). From (1.5) we now obtain, on applying $\mathrm{Hom}_A(\cdot\ , A)$, the exact sequence

$$\mathrm{Hom}_A(M,A) \to \hat{X} \to \hat{Y} \to \mathrm{Ext}_A^1(M,A) \to \mathrm{Ext}_A^1(X,A) \ ,$$

again with the extreme terms null. This yields the exact sequence under (i).

As $\mathrm{Ext}_A^2(M,A) = 0$, M being of projective dimension $\leqslant 1$, it follows now that $M \longmapsto M^*$ defines an exact functor. Finally one can easily check in terms of the exact sequences (1.5) and the dual sequences of (i) that the double dual map $M \to M^{**}$ is induced from the double dual maps $X \to \hat{\hat{X}}$, hence is also an isomorphism.

From the Proposition it follows that the maps

$$[M] \longmapsto \overline{[M]} = [M^*] \tag{1.7}$$

and

$$[X] \longmapsto \overline{[X]} = - [\hat{X}] \tag{1.8}$$

define involutory automorphisms $\bar{\ }$ of $K_oT(A)$, and of $K_o(A)$, respectively, the latter inducing an involutory automorphism of $\mathrm{Cl}(A)$.

Finally note that going to the limits for general linear groups we also get automorphisms of $K_1(A)$ and $K_1(A)$ induced by the involution.

1.3 Proposition: The automorphisms $^-$ defined above commute with the maps of Theorems 1 and 2.

This is indeed fairly obvious now, on the basis of Propositions 1.1 and 1.2, taking for granted the fact that localisation preserves the involution. The only "unexpected" result is the change in sign in the definition of $^-$ on $K_0(A)$ and $Cl(A)$. This reflects the interchange between X and Y in taking duals in the sequence (1.5). Of course on $\mathbb{Z}$ the involution is now $n \longmapsto \bar{n} = -n$.

Remark: Proposition 1.2 - appropriately reworded - remains true in the context of module theory over orders, without an involution being present. One then has to view $\hat{X}$, M^* as modules of the opposite hand from X, M. This however will not be needed here.

§2 Hermitian modules

Here we develop the basic notion of a Hermitian module and pose the discriminant problem. Throughout $(A, ^-)$ is an involution algebra (A is always finite dimensional, separable over F), and $\mathfrak{A}$ is an order with $\bar{\mathfrak{A}} = \mathfrak{A}$. Unless otherwise stated $\mathfrak{A}$-modules have $\mathfrak{A}$ acting on the right.

A Hermitian $\mathfrak{A}$-module is a pair (X,h) where X is locally free over $\mathfrak{A}$ (say of rank q) and h is a non-degenerate Hermitian form on the A-module $V = X \otimes_{\mathfrak{A}} A = X \otimes_{\mathcal{O}} F$. Thus $h : V \times V \to A$ is biadditive, A-linear in the second variable, i.e., $h(v_1, v_2 a) = h(v_1,v_2)a$ and so that $\overline{h(v_1,v_2)} = h(v_2,v_1)$ with the matrix $(h(v_i,v_j))$ lying in $GL_q(A)$.

Example: $A = F\Gamma$ is the group ring over F of a finite group Γ, and $\bar{\gamma} = \gamma^{-1}$ for all $\gamma \in \Gamma$. The language of Hermitian forms over $F\Gamma$ is then equivalent with that of Γ-invariant symmetric forms over F, (cf. [FMcl] §7) (see also Chapt. V §3).

If (X_i,h_i) $(i = 1,2)$ are Hermitian A-modules, then so is their orthogonal sum

$$(X_1,h_1) \perp (X_2,h_2) = (X_1 \oplus X_2, h_1 \perp h_2) .$$

Here $h_1 \perp h_2$ restricts to h_i on $X_i \times X_i$ and is null on $X_i \times X_j$, for $i \neq j$. We shall write $K_oH(A)$ for the Grothendieck group of (isometry classes of) Hermitian A-modules and orthogonal sums.

The general problem of change of ring will be discussed later. It is however immediately clear that transitions from the order to the algebra, or localisation yield homomorphisms on K_oH.

2.1 Proposition: The map on K_oH associated with change of ring is surjective in the following cases (i) $A \to A$, (ii) $A \to A_p$, (iii) $A \to A_p$ for any prime divisor p of F (in the global case).

Proof: In fact the actual result is stronger. Thus in case (ii), which lies deeper than the other two, it was proved in [F5] (Theorem 11.1) that given any A-module V, and a non-degenerate Hermitian form k on V_p over A_p, there exists a form h on V over A so that h_p is isometric to k. This immediately implies the assertion for this case.

Case (i) is entirely trivial. If we have a form h on the free A-module V just choose a free A-module X spanning V, to get (X,h). Finally the assertion for case (iii) follows from that for cases (i), (ii) and easy standard approximation.

Remark: In cases (ii) and (iii) the Proposition can be strengthened by considering simultaneously localisation at a finite set of prime divisors. This will however not be needed.

Let (X,h) be a Hermitian A-module, with X actually free, say of rank q. We have immediately

2.2 Proposition: The "discriminant matrix" $(h(x_i,x_j))$ for any free basis $\{x_i\}$ of X over A lies in $GL_q(A)$ and is symmetric under the matrix extension of $^-$. It is uniquely determined by (X,h) modulo a substitution by a matrix $\bar{a}(h(x_i,x_j))a$, where $a \in GL_q(A)$.

(An element x in a ring R with involution j is symmetric if $x^j = x$) .

Our aim is to use this Proposition to define a discriminant, just as for ordinary lattices (say over $\mathcal{O}$) - not in the weak classical sense, i.e., modulo "units" but modulo "unit squares" (both locally and globally) as in [F1] . The approach which first suggests itself is to use our general definition of the determinant and to define the discriminant of (X,h) to be the class - in the appropriate sense - of the map

$$\mathrm{Det}(h(x_i,x_j)) : \chi \longmapsto \mathrm{Det}_\chi(h(x_i,x_j)) .$$

Whenever however χ is symplectic (see below in §4 for the precise definition) then $\mathrm{Det}_\chi(h(x_i,x_j))$ is a square and we shall indeed see that it then has a canonical square root - just as in the special case of the quaternions discussed in the introduction. It would then however be extremely cumbersome to deal with the symplectic and the non-symplectic case separately, and indeed our notion of a Pfaffian will lead to a unified treatment.

§3. Pfaffians of matrices

We develop here the theory of Pfaffians for matrices - not as usual for skew symmetric matrices, but for matrices which are symmetric with respect to a symplectic involution. (See [F8], [F9] and [F10]).

In this section L can be any field of characteristic $\neq 2$. The matrix ring $M_n(L)$ acts from the left on the n-dimensional vector space L^n, thus viewed as a column space. We consider involutions of $M_n(L)$, fixing L elementwise. Every such involution j is the adjoint of a non-singular symmetric or skew symmetric bilinear form β on L^n, i.e. so that for all $x, y \in L_n$, all $P \in M_n(L)$,

$$\beta(x, Py) = \beta(P^j x, y) . \qquad (3.1)$$

j certainly determines the property of β being symmetric or skew symmetric, as we shall presently note. In the first case we call j _orthogonal_, in the second case _symplectic_. In this section j is to be a symplectic involution of $M_n(L)$. The corresponding forms will for short just be called skew forms. A matrix S with $S^j = S$ will be called j-symmetric. These j-symmetric matrices form an L-space of dimension $n(n-1)/2$. In the orthogonal case it would be $n(n+1)/2$.

3.1 _Proposition: Let_ $S \in GL_n(L)$ _be_ j-_symmetric. Then_

$$S = P^j P, \quad \text{for some} \quad P \in GL_n(L) .$$

_If_ (the identity matrix being denoted by I)

$$I = P_1^{\,j} P_1 , \qquad P_1 \in GL_n(L) ,$$

then $\det P_1 = 1$.

The proof of both assertions is well known. The first one is an immediate consequence of the fact that all skew forms of given dimension have a hyperbolic basis, hence are isometric.

For the second assertion one may assume L to be algebraically closed. One views L^n as $L[X]$-module, the indeterminate X acting via the matrix P_1 of the Proposition. One then shows that to every 1-dimensional eigen space V of X belonging to the eigenvalue λ there corresponds a X-subspace $V^{\perp}$ (its orthogonal complement) containing V, with codimension 1, so that $L^n/V^{\perp}$ belongs to λ^{-1}. This reduces the dimension to $n-2$.

By the proposition, the j-symmetric matrix S determines the determinant $\mathrm{Det}(P)$ for $S = P^jP$. We define its <u>Pfaffian</u> by

$$\mathrm{Pf}^j(S) = \mathrm{Det}(P) \quad (=\mathrm{Det}(P^j)) \ . \tag{3.2}$$

(Indeed determinants of matrices are invariant under involutions over the basefield.)

3.2 Proposition: $\mathrm{Pf}^j(I) = 1$. *If* $S \in GL_n(L)$ *is j-symmetric then*

$$\mathrm{Det}(S) = \mathrm{Pf}^j(S)^2 \ .$$

If $P \in GL_n(L)$ *then*

$$\mathrm{Pf}^j(P^jSP) = \mathrm{Pf}^j(S)\,\mathrm{Det}(P) \ .$$

We shall omit the proof for this, as for many other propositions in this section; they are mostly straightforward.

3.3 Proposition: *If* $S_i \in GL_{n_i}(L)$ *is* j_i*-symmetric,* j_i *the adjoint involution of a skew form* β_i $(i = 1,2)$, *then*

$$S = \begin{pmatrix} S_1 & 0 \\ 0 & S_2 \end{pmatrix} \in GL_n(L) \;, \quad n = n_1 + n_2 \;,$$

is j-symmetric, j the adjoint involution of the orthogonal sum $\beta_1 \perp \beta_2$. Also

$$Pf^j(S) = Pf^{j_1}(S_1)\, Pf^{j_2}(S_2) \;.$$

Next let

$$< , > : L^n \times L^n \to L$$

be a non-singular pairing. Let k be the antiautomorphism of $M_n(L)$ with

$$<P^k x, y> = <x, Py> \;.$$

We define the hyperbolic skew form κ on L^{2n} by

$$\kappa\left(\begin{pmatrix} v_1 \\ v_2 \end{pmatrix}, \begin{pmatrix} w_1 \\ w_2 \end{pmatrix}\right) = <v_1, w_2> - <w_1, v_2>$$

$v_i, w_i \in L^n$. Let j be its adjoint involution.

3.4 Proposition: Let $T \in GL_n(L)$. Then

$$S = \begin{pmatrix} T^k & 0 \\ 0 & T \end{pmatrix}$$

is j-symmetric, and

$$Pf^j(S) = Det(T) .$$

Next let $s: L \to K$ be an embedding of fields, extended in some way to a homomorphism

$$s: M_n(L) \to M_n(K) .$$

3.5 Proposition: Let j be a symplectic involution of $M_n(L)$, k one of $M_n(K)$ so that $j \circ s = s \circ k$. If $S \in GL_n(L)$ is j-symmetric, then S^s is k-symmetric, and

$$(Pf^j(S))^s = Pf^k(S^s) .$$

Corollary 1. Let ω be an automorphism of L , and j a symplectic involution of $M_n(L)$. Then the map $P \longmapsto P^k = P^{\omega^{-1}j\omega}$ is also a symplectic involution. If S is j-symmetric then S^ω is k-symmetric and

$$(Pf^j(S))^\omega = Pf^k(S^\omega) .$$

Next observe that we may extend the identity map on L by a conjugation in $M_n(L)$. We get

Corollary 2. *Let* j *be a symplectic involution of* $M_n(L)$, *and let* $R \in GL_n(L)$. *Then there is a unique involution* k, *also symplectic, with*

$$R^{-1} P^j R = (R^{-1} P R)^k .$$

If S *is* j*-symmetric then* $R^{-1} S R$ *is* k*-symmetric and*

$$Pf^j(S) = Pf^k (R^{-1} S R) .$$

Finally we shall give another result on change of involution, which also implies the last corollary.

3.6 Proposition: *Let* j *be a symplectic involution of* $M_n(L)$ *and* R *a* j*-symmetric matrix in* $GL_n(L)$. *Then* $P \mapsto P^k = R^{-1} P^j R$ *defines a symplectic involution of* $M_n(L)$ *and conversely every symplectic involution* k *is of this form. Moreover a matrix* S *is* k*-symmetric if and only if* RS *is* j*-symmetric, and then*

$$Pf^j(RS) = Pf^j(R)\ Pf^k(S) .$$

Proof: Given j, any other involution is easily shown to be of form $P^k = R^{-1} P^j R$ and one also sees immediately that k is an involution precisely when $R^j = cR$, $c \in L^*$. But then $c^2 = 1$, $c = \pm 1$. If $c = -1$, we get an orthogonal involution, if $c = 1$, a symplectic one. The symmetry condition is obvious.

Now let $S = P^k P$, so that $Pf^k(S) = Det(P)$, and let $R = T^j T$ so that $Pf^j(R) = Det(T)$. Then $RS = R.R^{-1}P^jR\ P = P^j\ T^j\ TP = (TP)^j\ (TP)$, hence $Pf^j(RS) = Det(TP) = Pf^j(R)\ Pf^k(S)$.

§4. Pfaffians of algebras

Based on the results of the last section we can now define a Pfaffian for involution algebras, (cf. [F8], [F9], [F10]) following the same philosophy that produced our determinants for algebras (cf. I. §2). This will lead to the definition of canonical square roots for determinants of symmetric elements at symplectic representations (see Proposition 4.5) and thus open the way for a good definition of the discriminant. We shall see (cf. Proposition 4.6) that the determinants of symmetric elements (at all representations) are determined by their Pfaffians (at the symplectic representations). Thus the Pfaffians indeed allow a unified approach to the discriminant problem, not requiring any distinction of cases.

As before A is a finite dimensional, separable F-algebra, and from now on $(A,\bar{\ })$ is an involution algebra over F (i.e., fixing F elementwise). F may be any field of characteristic $\neq 2$. A representation

$$T : (A,\bar{\ }) \to (M_n(F_c),\beta) \tag{4.1}$$

is given by a representation $A \to M_n(F_c)$, again, by abuse of notation, to be denoted by T and a symmetric or skew form β on $F_c^{\,n}$, so that if j is its adjoint involution, we have

$$T(\bar{a}) = T(a)^j, \quad \text{for all} \quad a \in A . \tag{4.2}$$

The extension (cf. I. (2.4)) $M_q(A) \to M_{nq}(F_c)$ will again satisfy (4.2), where $\bar{\ }$, j are now the matrix extensions of the given involutions (cf. (1.1)). If j is an orthogonal (symplectic) involution we shall call T an <u>orthogonal</u> (<u>symplectic</u>) <u>representation</u>. We shall in the sequel discuss everything in symplectic terms, this being the case we want. Analogous remarks also apply of course to the orthogonal case. (If $A = F\Gamma$ is the group ring of a finite group Γ, with

$\bar{\gamma} = \gamma^{-1}$ for all $\gamma \in \Gamma$ then T is orthogonal (symplectic) if, and only if, the representation $\Gamma \to GL_n(F_c)$ factorises through an orthogonal (symplectic) group).

Two symplectic representations (T,β), (T',β') (of degree n say) are <u>isometric</u> if there exists $R \in GL_n(F_c)$ with

$$\begin{cases} T'(a) = R^{-1}T(a)\,R \quad \text{for all} \quad a \in A \\ \beta'(x,y) = \beta(Rx,Ry) \quad \text{for all} \quad x,\, y \in F_c^{\,n} \,. \end{cases} \tag{4.3}$$

This implies for the adjoint involutions j of β and k of β' that

$$RP^k\,R^{-1} = (RPR^{-1})^j \quad \text{for all} \quad P \in M_n(F_c) \,. \tag{4.4}$$

The arrow in (4.1) cannot be interpreted as a map, as the objects on either side of it are of different types. Two forms β and β' will however have the same adjoint involution j if, and only if, $\beta' = a\beta$ for some nonzero scalar a. In this case the representations (T,β) and (T,β') will be isometric. We can thus equivalently speak of isometry classes of pairs (T,j).

The sum of two symplectic representations is defined by

$$(T_1,\beta_1) + (T_2,\beta_2) = \left(\begin{pmatrix} T_1 & 0 \\ 0 & T_2 \end{pmatrix}, \beta_1 \perp \beta_2 \right) , \tag{4.5}$$

where $\beta_1 \perp \beta_2$ is the orthogonal sum. The isometry classes form a semigroup under this addition, we denote its Grothendieck group by $K^s_{A,F}$ (not indicating explicitly the given involution $^-$). Forgetting the forms yields a homomorphism $\xi\colon K^s_{A,F} \to K_{A,F}$. We shall write $G^s_{A,F}$ for the classes of indecomposable representations (T,β).

4.1 Proposition: (i) $K^s_{A,F}$ is the free Abelian group on $G^s_{A,F}$

(ii) If $\chi \in G_{A,F}$, then either (a) there is a unique class in $G^s_{A,F}$ which maps onto χ under ξ and then $\chi = \bar{\chi}$, or (b) there exists a unique class in $G^s_{A,F}$ which maps onto $\chi + \bar{\chi}$ under ξ. Cases (a) and (b) are mutually exclusive.

(iii) There are no other classes in $G^s_{A,F}$.

Corollary: $\xi: K^s_{A,F} \to K_{A,F}$ is injective, and every element of $K^s_{A,F}$ which under ξ maps onto the actual class of a representation will correspond to a symplectic representation.

Remark: In view of the Corollary we shall view $K^s_{A,F}$ as a subgroup of $K_{A,F}$. An element χ of $\mathrm{Im}\,\xi$ will be called symplectic. By slight abuse of terminology we shall speak of the underlying representation $T: A \to M_n(F_c)$ of a symplectic representation (T,β) itself as symplectic, if we do not wish to specify β. As ξ is injective, T will certainly determine (T,β) to within isometry. Moreover, as we shall see, T itself determines the Pfaffian, without the knowledge of β.

For the proof of 4.1, and indeed for a proper understanding of $K^s_{A,F}$, we shall first give some further interpretations of $K^s_{A,F}$, and of $G^s_{A,F}$. Analogous remarks also apply with symplectic representations replaced by orthogonal ones. Firstly, in analogy to I. Proposition 2.3 and the remark following its proof, we get an isomorphism $K^s_{A,F} \cong K^s_{A \otimes F_c, F_c}$, the tensor product being taken over F, and the involution of A extending to that of $A \otimes F_c$, with F_c pointwise fixed. This isomorphism is induced by associating with a representation $T: A \to M_n(F_c)$ the compositum

$$A \otimes_F F_c \to M_n(F_c) \otimes_F F_c \to M_n(F_c) .$$

It maps $G^s_{A,F}$ bijectively onto $G^s_{A\otimes F_c,F_c}$. Next, in the obvious way, $K^s_{A\otimes F_c,F_c}$ can also be identified with the Grothendieck group of pairs (V,β) under orthogonal sum, where V is, say, a left $A\otimes F_c$-module, $\beta\colon V\times V\to F_c$ is a (non-singular) skew form and for all $a\in A\otimes F_c$, all $v_1, v_2\in V$

$$\beta(av_1, v_2) = \beta(v_1, \bar{a}v_2) .$$

Finally using the trace $t\colon A\otimes_F F_c\to F_c$ as an involution trace (cf. [FMc1] §7) we can associate with β a unique skew-Hermitian form $h_\beta\colon V\times V\to A\otimes F_c$ so that

$$\beta(v_1, v_2) = t\, h_\beta(v_1, v_2) .$$

The Grothendieck group $K^s_{A,F} = K^s_{A\otimes F_c,F_c}$ of pairs (V,β) as above thus becomes identified with that of non-singular skew Hermitian forms over $A\otimes_F F_c$ under orthogonal sum (see the paper [FMc1] quoted above). The elements of $G_{A,F} = G_{A\otimes F_c,F_c}$ then correspond to the unsplittable such forms. Moreover in this new interpretation of $K^s_{A,F}$ and correspondingly of $K_{A,F}$ (as $K_o(A\otimes_F F_c))$, the map $K^s_{A,F}\to K_{A,F}$ is just the forgetful map: form $\mapsto$ underlying module. One can thus in particular reinterpret the known results on forms in terms of representations, and the last Proposition 4.1 follows then from [Mc] (§2) and [FMc2] (4.3).

Given (4.3), (4.4), then by Corollary 2 to Proposition 3.5,

$$\mathrm{Pf}^k\, T'(a) = \mathrm{Pf}^j T(a) \quad \text{for } \bar{a} = a .$$

By the injectivity of ξ we conclude

4.2 Proposition: *Let* (T,β) *be a symplectic representation*, j *the adjoint involution of* β *and* $a = \bar{a} \in GL(A)$. *Then*

$$Pf^{j}T(a) = Pf_{\chi}(a)$$

only depends on the class χ *of* T *in* $K^{s}_{A,F}$.

The properties of the Pfaffian of $(A,\bar{\ })$, which the last proposition defines, now follow quickly from those of Pfaffian for matrices. (See [F8], where most of these results were stated first, [F9] and [F10]). Note that we now have a map

$$Pf(a): \chi \longmapsto Pf_{\chi}(a) \ , \quad a \text{ symmetric in } GL(A), \ \chi \text{ symplectic.}$$

4.3 Proposition: *Let* $a = \bar{a} \in GL_q(A)$. *The map* $Pf(a): \chi \longmapsto Pf_{\chi}(a)$ *extends to a homomorphism* $K^{s}_{A,F} \to F_c^{\ *}$ *of* Ω_F*-modules. Thus* $Pf: a \longmapsto Pf(a)$ *is a map of the set of symmetric elements of* $GL(A)$ *into* $Hom_{\Omega_F}(K^{s}_{A,F}, F_c^{\ *})$.

Proof: By Proposition 3.3 and Corollary 1 to Proposition 3.5.

4.4 Proposition: *If for* $i = 1,2,$

$$a_i = \overline{a_i} \in GL_{q_i}(A) \ ,$$

then $Pf\begin{pmatrix} a_1 & 0 \\ 0 & a_2 \end{pmatrix} = Pf(a_1)\,Pf(a_2)$, *i.e.*,

$$\mathrm{Pf}_\chi \begin{pmatrix} a_1 & 0 \\ 0 & a_2 \end{pmatrix} = \mathrm{Pf}_\chi(a_1)\ \mathrm{Pf}_\chi(a_2)\ .$$

Proof: By Proposition 3.3.

Let Det^s be the restriction of Det to $K^s_{A,F}$.

4.5 Proposition: Let $a = \bar{a} \in GL_q(A)$. Then

$$\mathrm{Det}^s(a) = \mathrm{Pf}(a)^2$$

i.e.,

$$\mathrm{Det}_\chi(a) = \mathrm{Pf}_\chi(a)^2\ , \qquad \chi \in K^s_{A,F}\ .$$

Let moreover $b \in GL_q(A)$. Then

$$\mathrm{Pf}(\bar{b}\ a\ b) = \mathrm{Det}^s(b)\ \mathrm{Pf}(a)$$

i.e.,

$$\mathrm{Pf}_\chi(\bar{b}\ a\ b) = \mathrm{Det}_\chi(b)\ \mathrm{Pf}_\chi(a)\ .$$

Proof: By Proposition 3.2.

We now see that Pf(a) is the square root of $\mathrm{Det}^s(a)$.

Next recall the definition of $^-$ on $K_{A,F}$ (cf. (1.2)). Write, for $\chi \in K_{A,F}$,

$$\mathrm{Tr}(\chi) = \chi + \bar{\chi} . \tag{4.6}$$

By Proposition 4.1, Tr is an Ω_F homomorphism $K_{A,F} \to K^s_{A,F}$, giving rise to a homomorphism

$$\mathrm{Nr}\colon \mathrm{Hom}_{\Omega_F}(K^s_{A,F}, X) \to \mathrm{Hom}_{\Omega_F}(K_{A,F}, X) \tag{4.7}$$

for any Ω_F-module X , defined by

$$(\mathrm{Nr}f)(\chi) = f(\mathrm{Tr}(\chi))$$

4.6 <u>Proposition</u>: <u>Let</u> $a = \bar{a} \in GL_q(A)$. <u>Then</u> $\mathrm{Nr}(\mathrm{Pf}(a)) = \mathrm{Det}(a)$, i.e., <u>for all</u> $\chi \in K_{A,F}$

$$\mathrm{Pf}_{\chi+\bar{\chi}}(a) = \mathrm{Det}_\chi(a) .$$

<u>Proof</u>: This follows from Proposition 3.4.

Note that, by 4.6, the determinant of a symmetric element at all χ is determined by its Pfaffian at the symplectic χ .

<u>Examples</u>: (a) If E is a field of characteristic other than 2 ,then

$$\begin{pmatrix} a & b \\ c & d \end{pmatrix}^{\dagger} = \begin{pmatrix} d & -b \\ -c & a \end{pmatrix} \tag{4.8}$$

defines a symplectic involution of $M_2(E)$, the adjoint of the "alternating hyperbolic plane". The only symmetric elements are the scalar matrices

$$aI = \begin{pmatrix} a & 0 \\ 0 & a \end{pmatrix} .$$

We shall show that indeed

$$Pf^{\dagger}(aI) = a . \tag{4.9}$$

We only have to note that e.g.

$$\begin{pmatrix} 0 & 1 \\ -a & 0 \end{pmatrix} \begin{pmatrix} 0 & 1 \\ -a & 0 \end{pmatrix}^{\dagger} = aI$$

and that

$$\mathrm{Det} \begin{pmatrix} 0 & 1 \\ -a & 0 \end{pmatrix} = a .$$

Alternatively we could have used the matrix $\begin{pmatrix} a & 0 \\ 0 & 1 \end{pmatrix}$.

Now suppose that $E = K_c$ is the algebraic closure of a field K and that D is a quaternion algebra over K with maximal subfield L . Denote the generating automorphism of the quadratic extension L of K by $x \longmapsto x'$. Then the (to within equivalence) unique irreducible representations T of D over K embeds D in $M_2(E)$, the elements of $T(D)$ being the matrices

$$\begin{pmatrix} x & y \\ cy' & x' \end{pmatrix}$$

for variable $x, y \in L$ and with c fixed in K^*. D is a division algebra if and only if c is not a norm from L. Moreover the standard involution $z \mapsto \bar{z}$ of D, i.e., that involution, for which $z\bar{z} = \mathrm{nrd}(z)$ is the reduced norm, is just the restriction of † (cf. (4.8)) under the embedding T. It follows that if χ denotes the class of T, then

$$\mathrm{Pf}_\chi(a\ 1_D) = a \quad \text{for} \quad a \in K^* . \tag{4.10}$$

More generally let $M_m(D)$ be the involution algebra with the matrix extension of $^-$ as involution. Then any symmetric element is of the form

$$b = c(\mathrm{diag}(a_i\ 1_D))\bar{c}\ , \quad c \in GL_m(D), \quad a_i \in K^*\ , \tag{4.11}$$

where $\mathrm{diag}(a_i\ 1_D)$ is the diagonal matrix with entries $a_i\ 1_D$. If now χ is the class of the irreducible representations of $M_m(D)$ we get

$$\mathrm{Pf}_\chi(\mathrm{diag}(a_i\ 1_D)) = \Pi\ a_i\ ,$$

(this is essentially Wall's Pfaffian (cf. [Wa3])) and more generally

$$\mathrm{Pf}_{n\chi}(\mathrm{diag}(a_i\ 1_D)) = \Pi\ a_i^n\ . \tag{4.12}$$

Finally take $K = \mathbb{R}$ the field of real numbers and $D = \mathbb{H}$ the

Hamiltonian quaternion. Then we may take the a_i to have values ± 1 only, and we get

$$\mathrm{Pf}_\chi(\mathrm{diag}(a_i\, 1_D)) = (-1)^q$$

q = number of negatives entries,

or in other words

$$\mathrm{Pf}_\chi(\mathrm{diag}(a_i\, 1_D)) = (-1)^{(m-s)/2}\,, \tag{4.13}$$

s the signature over $\mathbb{H}$.

(b) We consider a second example, the hyperbolic Hermitian plane over an involution algebra $(A,\bar{\ })$. The underlying module is $A^2 = A \times A$ with basis $x = (1,0)$ and $y = (0,1)$ and h is given by

$$h(x,x) = 0 = h(y,y), \qquad h(x,y) = 1 = h(y,x)\ .$$

The discriminant matrix is

$$e = \begin{pmatrix} 0 & 1_A \\ 1_A & 0 \end{pmatrix} \in GL_2(A)\ . \tag{4.14}$$

In $GL_2(A \otimes_F F_c)$ we have

$$e = f\bar{f} \;, \qquad f = \sqrt{2}^{\,-1}\begin{pmatrix} 1_A & -\sqrt{-1}\,1_A \\ 1_A & \sqrt{-1}\,1_A \end{pmatrix} , \tag{4.15}$$

where $\bar{\ }$ is extended to $A \otimes_F F_c$ via the first tensor factor. If T is a symplectic representation of A, extended to $A \otimes_F F_c$ - and as such again symplectic, then by Proposition 4.5,

$$\begin{aligned} Pf_\chi(e) &= Det_\chi(f) \\ &= Det\ \sqrt{2}^{\,-1}\begin{pmatrix} T(1_A) & -\sqrt{-1}\,T(1_A) \\ T(1_A) & \sqrt{-1}\,T(1_A) \end{pmatrix} = (\sqrt{-1})^{\deg(\chi)} , \end{aligned}$$

where $\deg(\chi)$ is the degree of χ, i.e., the order of T. Hence as $\deg(\chi)$ is even for symplectic χ,

$$Pf_\chi(e) = (-1)^{\deg(\chi)/2} . \tag{4.16}$$

§5. Discriminants and the Hermitian classgroup

We shall now introduce the Hermitian classgroup, this to be the group in which discriminants take their values. Throughout $\bar{A} = A$. For a first informal discussion suppose that $\mathcal{O}$ is a complete local ring. If (X,h) is a Hermitian A-module and $\{x_i\}$ a basis of X over A then by Proposition 2.2 and by Propositions 4.3 and 4.5 the element $Pf(h(x_i,x_j))$ (the Pfaffian of discriminant matrix) belongs to $Hom_{\Omega_F}(K^s_{A,F}, F_c^*)$, uniquely determined modulo $Det^s A^*$. In other words, the obvious choice for the Hermitian classgroup is the quotient $Hom_{\Omega_F}(K^s_{A,F}, F_c^*)/Det^s A^*$, and indeed we shall see that in this local case the Hermitian classgroup is isomorphic to this quotient - although not in the first place defined as such. Going over to the case when $\mathcal{O}$ is the ring of algebraic integers in a number field F ("$\mathcal{O}$ is global" as we shall say) one would then piece the local discriminants together to take values in the group $\mathrm{Hom}_{\Omega_F}(K^s_{A,F}, J(F_c))/\mathrm{Det}^s\, U\mathcal{A}$. This group will indeed play a role, but this definition is still not strong enough. To take it as the Hermitian classgroup would lose us non trivial global information. Thus, e.g., there is no natural map from this group onto $\mathrm{Cl}(\mathcal{A})$, but there is one from the global Hermitian classgroup, as we shall define it, which reflects the functor $(X,h) \longmapsto X$. Moreover, as we shall see in III, the map from the global group to the above group (a restricted product of local groups) may have both non-trivial kernel and non-trivial cokernel.

We shall specifically be interested in three cases and we shall concentrate on these: (i) $\mathcal{O}$ is global - i.e., the ring of integers in a number field, (ii) $\mathcal{O}$ is local - meaning $\mathcal{O}$ is the ring of integers in a finite extension of $\mathbb{Q}_p$, (iii) $\mathcal{O} = F$. The formulations to be given will differ in these three cases. In order to show that they have a common source we shall, in an appendix to this section, outline a unified treatment of basic concepts, which is sufficiently general to cover the case when $\mathcal{O}$ is any Dedekind domain. Apart from this however, it will be understood for the remainder of this chapter II and for chapter III that $\mathcal{O}$ and F are as given above under (i), (ii) or (iii). Many of our results will of course be more general.

In the sequel JA is the idele group, $U\mathcal{A}$ the group of unit ideles ($\mathcal{O}$ global) defined in I §2. We consider the group

$$G(\mathcal{A}) = \begin{cases} (\mathrm{Det}\, A^*/\mathrm{Det}\, \mathcal{A}^*) \times \mathrm{Hom}_{\Omega_F}(K^s_{A,F}, F_c^*), & \mathcal{O} \text{ local}, \\ (\mathrm{Det}\, JA/\mathrm{Det}\, U\mathcal{A}) \times \mathrm{Hom}_{\Omega_F}(K^s_{A,F}, F_c^*), & \mathcal{O} \text{ global}, \\ \mathrm{Hom}_{\Omega_F}(K^s_{A,F}, F_c^*)\ , & \mathcal{O} = F \ \text{ i.e.,} \quad \mathcal{A} = A\ , \end{cases} \tag{5.1}$$

and we define a homomorphism

$$\Delta\colon \operatorname{Det} A^* \to G(A) \tag{5.2}$$

by

$$\Delta(g) = \begin{cases} (g^{-1} \bmod \operatorname{Det} A^*,\ g^s)\ , & \mathcal{O} \text{ local,} \\ (g^{-1} \bmod \operatorname{Det} UA,\ g^s)\ , & \mathcal{O} \text{ global,} \\ g^s & ,\ \mathcal{O} = F\ , \end{cases}$$

where g^s is the restriction of g to $K^s_{A,F}$. The Hermitian class-group is now defined by

$$\mathrm{HCl}(A) = \operatorname{Cok} \Delta\ . \tag{5.3}$$

It is more suggestive to write it, e.g., in the case $\mathcal{O}$ global, as

$$\mathrm{HCl}(A) = ((\operatorname{Det} JA/\operatorname{Det} UA) \times \operatorname{Hom}_{\Omega_F}(K^s_{A,F}, F_c^*))/\operatorname{Det} A^*\ , \tag{5.4}$$

viewing it as a set of orbits in $G(A)$ under the action of $\operatorname{Det} A^*$.

Next we define, with $J(F_c)$ again the idele group,

$$\mathrm{HCl}'(A) = \begin{cases} \operatorname{Hom}_{\Omega_F}(K^s_{A,F}, F_c^*)/\operatorname{Det}^s A^*, & \mathcal{O} \text{ local or } \mathcal{O} = F\ , \\ \operatorname{Hom}_{\Omega_F}(K^s_{A,F}, J(F_c))/\operatorname{Det}^s UA, & \mathcal{O} \text{ global}\ . \end{cases} \tag{5.5}$$

This is the group mentioned in the introduction to this section. Clearly

$$HCl'(A) = HCl(A) \quad \text{when} \quad \mathcal{O} = F\ . \tag{5.6}$$

In the other cases we have a homomorphism

$$\underline{P} = \underline{P}_A : HCl(A) \to HCl'(A)\ , \tag{5.7}$$

induced by the map $\underline{P}' : G(A) \to HCl'(A)$ with

$$\left.\begin{array}{l}(h \bmod \mathrm{Det}\, A^*,\ f) \\ \\ (h \bmod \mathrm{Det}\, UA,\ f)\end{array}\right\} \longmapsto h^s f \quad \left\{\begin{array}{ll}\bmod \mathrm{Det}^s\, A^*, & \mathcal{O} \text{ local,} \\ \\ \bmod \mathrm{Det}^s\, UA, & \mathcal{O} \text{ global.}\end{array}\right.$$

5.1 Proposition: *If $\mathcal{O}$ is local $\underline{P}$ is an isomorphism.*

Proof: It is clear that $\underline{P}'$ and hence $\underline{P}$ are surjective. Suppose $(h \bmod \mathrm{Det}\, A^*, f) \in \mathrm{Ker}\, \underline{P}'$. Then $h^s.f = u^s$, $u \in \mathrm{Det}\, A^*$, whence we may in fact suppose that $h^s.f = 1$. But then clearly $(h \bmod \mathrm{Det}\, A^*, f) = \Delta(h^{-1})$. Thus the class of this element in $HCl(A)$ is 1 . We have shown that $\mathrm{Ker}\, \underline{P} = 1$.

For $\mathcal{O}$ local we shall from now on identify $HCl(A) = HCl'(A)$ and use for it the description by the right hand side of (5.5).

When $\mathcal{O}$ is global and $\underline{P}$ is not necessarily an isomorphism (see III §2), we call $HCl'(A)$ the adelic Hermitian classgroup, writing

$$\mathrm{Ad}\, HCl(A) = HCl'(A)\ . \tag{5.8}$$

In this case we get, for each prime divisor p , a componentwise localisation map $G(A) \to G(A_p)$, using I , Propositions 2.2 and 2.3 to get e.g. from

$$\mathrm{Hom}_{\Omega_F}(K^s_{A,F}, (F_c \otimes_F F_p)^*) \quad \text{to} \quad \mathrm{Hom}_{\Omega_{F_p}}(K^s_{A_p,F_p}, F^*_{p,c}) .$$

For p infinite the map on the first component is of course just $(\mathrm{Det}\, JA/\mathrm{Det}\, UA) \to 1$. Going over to the classgroups we now have localisation maps

$$\begin{aligned} \lambda_p &: \mathrm{HCl}(A) \to \mathrm{HCl}(A_p) , \\ \lambda'_p &: \mathrm{Ad\,HCl}(A) \to \mathrm{HCl}'(A_p) = \mathrm{HCl}(A_p) . \end{aligned} \tag{5.9}$$

The map λ'_p is induced by the localisation on ideles.

Let $Y(F_{p,c})$ be the group of units of the integral closure of o_p in $F_{p,c}$. Write

$$\mathrm{HCl}''(A_p) = \mathrm{Hom}_{\Omega_{F_p}}(K^s_{A_p,F_p}, Y(F_{p,c}))/\mathrm{Det}^s\, A^*_p , \tag{5.10}$$

where of course $\mathrm{Det}^s\, A^*_p$ is indeed a subgroup of this Hom group. By (5.5) we can view $\mathrm{HCl}''(A_p)$ as a subgroup of $\mathrm{HCl}(A_p) = \mathrm{HCl}'(A_p)$.

5.2 Proposition: Let o be global.

(i) For all p, $\lambda_p = \lambda'_p \circ \underline{P}_A$.

(ii) The maps λ'_p set up an isomorphism

$$\mathrm{Ad\,HCl}(A) \cong \Pi(\mathrm{HCl}(A_p) \mid \mathrm{HCl}''(A_p))$$

(relative restricted product) .

As usual, given groups G_i with normal subgroups H_i, the relative restricted product $\Pi(G_i|H_i) \subset \Pi G_i$ is the inverse image of $\coprod(G_i/H_i)$ under the map $\Pi G_i \to \Pi(G_i/H_i)$. In this definition we need only know the H_i for all but a finite number of indices i.

The proposition follows almost immediately from the definitions.

We get a second description for $\mathrm{HCl}(A)$ in the global case, replacing in (5.4) the groups $\mathrm{Det}\, JA$ and $\mathrm{Det}\, A^*$ by $\mathrm{Hom}_{\Omega_F}(K_{A,F}, J(F_c))$ and $\mathrm{Hom}_{\Omega_F}(K_{A,F}, F_c^*)$ respectively. More precisely we have

5.3 Proposition: *Let*

$$\Delta': \mathrm{Hom}_{\Omega_F}(K_{A,F}, F_c^*) \to (\mathrm{Hom}_{\Omega_F}(K_{A,F}, J(F_c))/\mathrm{Det}\, UA) \times \mathrm{Hom}_{\Omega_F}(K_{A,F}^s, F_c^*)$$

be given by

$$\Delta'(g) = (g^{-1} \bmod \mathrm{Det}\, UA,\ g^s) .$$

Then the embedding $\mathrm{Det}\, JA \to \mathrm{Hom}_{\Omega_F}(K_{A,F}, J(F_c))$ *induces an isomorphism*

$$\mathrm{HCl}(A) \cong \mathrm{Cok}\, \Delta' .$$

The proof is the same as that for the analogous result on $\mathrm{Cl}(A)$ (Theorem 2. (vii)).

Next, still with F a number field, we write

$$\mathrm{Ad}\ \mathrm{HCl}(A) = \mathrm{Hom}_{\Omega_F}(K_{A,F}^s, J(F_c)/\mathrm{Det}^s JA . \tag{5.11}$$

The map $F_c^* \to J(F_c)$ then gives rise to a homomorphism

$$\underline{P}_A : \mathrm{HCl}(A) \to \mathrm{Ad}\ \mathrm{HCl}(A) . \tag{5.12}$$

Note that the use of the symbol Ad HCl(A) in (5.11), (5.12) is not strictly consistent with that in (5.8), but there should be no confusion. Moreover the $\underline{P}_A$ in (5.12) is not a special case of the map $\underline{P}_{\mathcal{A}}$ in (5.7), but the latter will only occur when $\mathcal{A}$ is an order, i.e. $\mathcal{O} \neq F$.

Now write for all finite p

$$\mathrm{HCl}''(A_p) = \mathrm{Im}\ [\mathrm{Hom}_{\Omega_{F_p}}(K^s_{A_p,F_p}, Y(F_{p,c})) \to \mathrm{HCl}(A_p)] . \tag{5.13}$$

As in the case of orders, we get again maps from both HCl(A) and Ad HCl(A) into $\mathrm{HCl}(A_p)$, both for p finite and infinite, and we have

5.4 <u>Proposition</u>: (i) <u>The diagrams</u>

$$\begin{array}{ccc} \mathrm{HCl}(A) & \longrightarrow & \mathrm{Ad}\ \mathrm{HCl}(A) \\ & \searrow \quad \swarrow & \\ & \mathrm{HCl}(A_p) & \end{array}$$

<u>commute and localisation sets up an isomorphism</u>

$$\mathrm{Ad}\ \mathrm{HCl}(A) \cong \prod (\mathrm{HCl}(A_p) \mid \mathrm{HCl}''(A_p)) ,$$

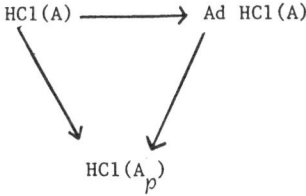

(product over all p.)

(ii) $\quad \mathrm{Cok}\ \underline{P}_{A} = \mathrm{Cok}\ \underline{P}_{A}$.

Proof: (i) Analogous to the proof of Proposition 5.2.

(ii) Both Cokernels are clearly

$$\mathrm{Hom}_{\Omega_F}(K^s_{A,F}, J(F_c))/[\mathrm{Hom}_{\Omega_F}(K^s_{A,F}, F_c^*).\mathrm{Det}^s JA] \quad .$$

We can now at last define the discriminant. Let (X,h) be a Hermitian A-module, X of rank q, say. Choose an A-basis $\{v_i\}$ of $X \otimes_A A$ and denote by V the (free)A-module generated by the v_i. Then $V \otimes_A A = X \otimes_A A$. If o is local, let $a \in GL_q(A) = \mathrm{Aut}_A(X \otimes_A A)$ be an element with $a V = X$. If o is global let similarly $a \in GL_q(A \otimes_F \mathrm{Ad}(F))$ be an element with $a_p V_p = X_p$ for all p. If $o = F$, neither X nor a enter into our definition.

Theorem 3. The image

$$d((X,h)) \in HCl(A)$$

of the element of $G(A)$, given by

$$(\mathrm{Det}(a) \bmod \mathrm{Det}\ A^*,\ \mathrm{Pf}(h(v_i,v_j)))\ , \quad o\ \text{local}\ ,$$

$$(\mathrm{Det}(a) \bmod \mathrm{Det}\ UA,\ \mathrm{Pf}(h(v_i,v_j)))\ , \quad o\ \text{global}\ ,$$

$$\mathrm{Pf}(h(v_i,v_j))\ , \quad o = F\ ,$$

<u>only depends on</u> (X,h) . d <u>defines a homomorphism</u> $K_0H(A) \to HCl(A)$. d is called the <u>discriminant</u>.

When $\mathcal{O}$ is local, then with the identification $HCl(A) = HCl'(A)$ via $\underline{P}_A$ we may view $d((X,h))$ to be represented by $Pf(h(x_i,x_j))$, $\{x_i\}$ a basis of X . In the case $\mathcal{O}$ global we have the

<u>Corollary</u>: <u>$Pd((X,h))$ is represented by the element with local components</u> $Pf(h(x_{i,p}, x_{j,p}))$, $\{x_{i,p}\}$ <u>a basis of</u> X_p <u>over</u> A_p .

<u>Proof of Theorem 3.</u> (Say for $\mathcal{O}$ global). An isometry $(X,h) \cong (X',h')$ leads with the right choice of bases to the same representative. On the other hand for given (X,h) and given $\{v_i\}$ we can change a by an element u with $u_p \in GL_q(A_p)$ for all p . We replace then $Det(a)$ by $Det(a)\,Det(u)$, $Det(u) \in Det\,UA$. On the other hand a change of basis $\{v_i\}$, via a matrix $c \in GL_q(A)$ replaces $Det(a)$ by $Det(a)Det(c)^{-1}$ and by Proposition 4.5 multiplies $Pf(h(v_i,v_j))$ by $Det^s(c)$. This yields the uniqueness. It remains to be shown that

$$d((X_1,h_1) \perp (X_2,h_2)) = d((X_1,h_1))\, d((X_2,h_2)) .$$

This however follows from Proposition 4.4.

<u>Remark</u>: In the local case we get a homomorphism

$$HCl(A) \to Hom(K^s_{A,F}, I(F_c)) , \tag{5.14}$$

where $I(F_c)$ is the group of fractional ideals of F_c , i.e., $I(F_c) = F_c^*/Y(F_c)$, $Y(F_c)$ the units of the ring of integers in F_c . This map is induced by the quotient map $F_c^* \to I(F_c)$. Similarly for

$\mathfrak{o}$ global we get a homomorphism

$$\mathrm{Ad\ HCl}(A) \to \mathrm{Hom}(K^s_{A,F}, I(F_c)) , \tag{5.15}$$

where again $I(F_c)$ is the group of fractional ideals, i.e., now $I(F_c) = J(F_c)/U(\mathfrak{o}_{F_c})$, U the unit ideles. We then obtain "discriminants", which are images of $\underline{P}_A d((X,h))$ under (5.15) (or locally (5.14)). This is a generalisation of the discriminant ideal in the classical theory of quadratic lattices. It plays a role in applications to number theory (cf. [F8]). For another generalisation see the next section.

Examples: (a) Our first example is A itself viewed as a Hermitian module via multiplication, i.e., $m_A(a_1,a_2) = \bar{a}_1 a_2$ for $a_1, a_2 \in A$. More generally $A^n = A \perp \dots \perp A$ then defines a Hermitian module.

(b) Next we consider the hyperbolic space $H(Y)$ on a locally free A-module Y - on the algebra level this has already been considered in example (b) of §4. The module underlying $H(Y)$ is $Y \times \hat{Y}$, $\hat{Y}$ the A-dual of Y , as defined in §1, see in particular (1.4). The form is then given, in terms of the pairing

$$< , > : Y \times \hat{Y} \to A$$

by

$$h((y_1,z_1), (y_2,z_2)) = < z_1,y_2 > + \overline{< z_2,y_1 >}$$

$(y_i \in Y, z_i \in \hat{Y})$, and extended to $(Y \times \hat{Y}) \otimes_A A$.

5.5 Proposition: With definitions (a), (b),

$$d((A,m_A)) = 1\,, \quad d(H(A^2)) = 1\,,$$

hence, for all n,

$$d((A^n,m_A^n)) = 1\,, \quad d(H(A^{2n})) = 1\,.$$

Proof: Choose as basis for the underlying A-module the canonical basis for the given Hermitian A-module, i.e., 1 in the first example, $(1,\hat{1})$ in the second. It then only remains to evaluate the Pfaffians. But $Pf(1) = 1$, and this gives $d((A,m_A)) = 1$. For $H(A^2)$ use formula (4.16).

In view of the last proposition, it makes sense to introduce a reduced Grothendieck group $\widetilde{K_oH(A)}$ as the kernel of the rank map

$$((X,h)) \longmapsto r_A(X) = r(X)\,,$$

i.e. given by the exact sequence

$$0 \to \widetilde{K_oH(A)} \to K_oH(A) \xrightarrow{r} \mathbb{Z} \to 0\,. \tag{5.16}$$

5.6 Proposition: (i) The sequence (5.16) splits naturally. More precisely, the maps

$$n \longmapsto ((A^n,m_A^n)) \quad (n > 0)$$

and

$$c\colon ((X,h)) \longmapsto ((X,h)) - r(X)((A,m_A))$$

give rise to an exact sequence

$$0 \to \mathbb{Z} \to K_oH(A) \xrightarrow{c} \widetilde{K_oH(A)} \to 0 . \tag{5.17}$$

(ii) $d: K_oH(A) \to HCl(A)$ factorises through c .

This is obvious, using for (ii) the formula of the preceding Proposition 5.5.

Remark: In a sense, $\widetilde{K_oH(A)}$ appears here as the Hermitian analogue to $Cl(A)$. Our (5.16) and (5.17) correspond to (1.1) and (1.2), respectively, in Chapter I. From the point of view of the Hom-description, however, the analogue to $Cl(A)$ is of course our Hermitian classgroup.

For completeness sake we mention that we also have an analogue to the classical Wittgroup. Let $Hyp(A)$ be the subgroup of $K_oH(A)$ generated by the classes of hyperbolic spaces $H(Y)$, as defined prior to Proposition 5.5. Thus $Hyp(A)$ is the image of the homomorphism $Hyp: K_o(A) \to K_oH(A)$ given by the hyperbolic functor. The Wittgroup is then

$$WH(A) = \text{Cok } Hyp = K_oH(A)/Hyp(A) . \tag{5.18}$$

Write $(-1)^{\deg/2}$ for the element of $\mathrm{Hom}_{\Omega_F}(K^s_{A,F}, F^*_c)$ with value $(-1)^{\deg(\chi)/2}$ at χ , and use the same symbol for its image in $HCl(A)$ (o local or $o = F$) , and for its image in $\mathrm{Ad}\ HCl(A)$ (o global) . We have then

5.7 Proposition: $\underline{P}_A d(H(Y)) = ((-1)^{\deg/2})^{r(Y)}$.

Proof: (For o global). Let $r(Y) = n$ and $Y = bA^n$ in A^n. Then, by Propositions 1.1 and 1.3, $\hat{Y} = \bar{b}^{-1}A^n$. By Theorem 3, $\underline{P}_A d(H(Y))$ is thus represented by $Pf(e)^n\, Det^s(b\bar{b}^{-1})$, with e as in (4.14). By (4.16), $Pf_\chi(e)^n = (-1)^{n\,\deg(\chi)/2}$. Also for $\chi \in R_\Gamma^s$, and so $\chi = \bar{\chi}$, we get

$$Det_\chi(b\bar{b}^{-1}) = Det_\chi(b)\, Det_{\bar{\chi}}(b)^{-1} = 1 .$$

This then yields the assertion.

One can now, just as in the ordinary theory of quadratic forms, define an adjusted discriminant d' by

$$d'((X,h)) = ((-1)^{\deg\chi/2})^{r(X)(r(X)-1)/2}\, d((X,h)) . \tag{5.19}$$

Then one sees easily that $\underline{P}_A d'$ defines a homomorphism on the subgroup of $K_0H(A)$, of modules with even rank, which has $Hyp(A)$ in its kernel.

The classical case: We briefly return to the situation which was our starting point for the discriminant problem, i.e., $A = F$, $A = o$ and trivial involution: $\bar{a} = a$. We shall show that we regain the classical definition. Indeed now $K_{A,F} = \mathbb{Z}\chi_1$, $\langle\chi_1\rangle = G_{A,F}$; here χ_1 is just the embedding $F \to F_c$. Thus $K_{A,F}^s = 2\,K_{A,F} = \mathbb{Z}\,(2\chi_1)$. Evaluation $f \longmapsto f(2\chi_1)$ yields an isomorphism $Hom_{\Omega_F}(K_{A,F}^s, F_c^*) \cong F^*$, and this maps, for any $a \in GL_q(o)$, $Det(a)$ into $Det_{\chi_1}(a)^2 = Det_{2\chi_1}(a)$. Therefore for o local or $o = F$,

$$HCl(o) \cong F^*/o^{*2} ,$$

and for any symmetric element $h \in GL_q(F)$, $Det(h) = Det_{\chi_1}(h) = Pf_{2\chi_1}(h)$. Thus indeed we do obtain the discriminant. For $\mathfrak{o}$ global we get the corresponding description for Ad $HCl(\mathfrak{o})$, putting the local pieces together. This example will be generalised in III. §2.

Appendix: Here $\mathfrak{o}$ is any Dedekind domain, F its quotient field (of characteristic $\neq 2$). We once and for all fix a set of prime divisors (equivalence classes of valuations) of F which includes all those coming from maximal ideals of $\mathfrak{o}$ - the "finite prime divisors", and finitely many others - the "infinite prime divisors". In the case $\mathfrak{o}$ global we let this be the usual set, in the case $\mathfrak{o}$ local it should be the set consisting of the divisor associated with the maximal ideal alone. For $\mathfrak{o} = F$ we get the trivial valuation of F, associated with the null ideal, under which F is its own completion. With F_p, $\mathfrak{o}_p$ the completions at p - with the convention however that for infinite p we set $\mathfrak{o}_p = F_p$, we define idele groups and groups of unit ideles in the usual manner. Now we do indeed have a general definition for $HCl(A)$ and $HCl'(A)$ as well as for $\underline{P}_A$ and for the discriminant d which looks exactly like that given for $\mathfrak{o}$ global, and this general definition leads then easily to the explicit formulation we have given for $\mathfrak{o}$ local and for $\mathfrak{o} = F$.

The reader will have noticed that in the global case we have used the notation $J(F_c)$ rather than the equally correct $J(\mathbb{Q}_c)$. It is the potentially more general validity of our definitions and of some of our results which leads us to adapt this type of suggestive convention. Although we are not giving general statements or proofs, the reader who is so minded can verify that much of what we are doing extends to more general Dedekind domains, using the definition of $J(F)$ - and hence of $J(F_c)$ - indicated above. At least we are not putting unnecessary notational obstacles in his way!

§6. Some homomorphisms

We shall derive some maps and exact sequences involving $HCl(A)$ and relating it to other groups which have arisen, in particular the group $K_0T(\mathcal{A})$.

We first define a map

$$\Phi : \mathrm{HCl}(\mathcal{A}) \to \mathrm{HCl}(A) \tag{6.1}$$

associated with the embedding $\mathcal{A} \subset A$. It can of course be described in terms of the general definition, given at the end of §5. Here and in similar places in the sequel we shall prefer to give the explicit description in terms of the special definitions for $\mathcal{O}$ global or local, respectively. For $\mathcal{O}$ local Φ is just the quotient map

$$\mathrm{Hom}_{\Omega_F}(K^s_{A,F}, F^*_c)/\mathrm{Det}^s\mathcal{A}^* \to \mathrm{Hom}_{\Omega_F}(K^s_{A,F}, F^*_c)/\mathrm{Det}^s A^* , \tag{6.2}$$

while for $\mathcal{O}$ global it is induced from the component projection

$$(\mathrm{Det}\ JA/\mathrm{Det}\ U\mathcal{A}) \times \mathrm{Hom}_{\Omega_F}(K^s_{A,F}, F^*_c)) \to \mathrm{Hom}_{\Omega_F}(K^s_{A,F}, F^*_c) . \tag{6.3}$$

We also have for $\mathcal{O}$ global the obvious quotient map

$$\mathrm{Ad}\ \Phi : \mathrm{Ad}\ \mathrm{HCl}(\mathcal{A}) \to \mathrm{Ad}\ \mathrm{HCl}(A) , \tag{6.4}$$

which may equivalently be viewed as a "relative restricted product" of the maps $\Phi_{\mathfrak{p}}$ of (6.1).

Next let

$$\rho_s : \mathrm{Hom}_{\Omega_F}(K_{A,F}, \cdot) \to \mathrm{Hom}_{\Omega_F}(K^s_{A,F}, \cdot) \tag{6.5}$$

be the restriction map. We now define a homomorphism

$$T : K_oT(A) \to HCl(A) \ . \tag{6.6}$$

When $\mathcal{O}$ is global we use the description in Theorem 2 (v); T is then just induced by the embedding of a direct factor

$$\mathrm{Det}\, JA/\mathrm{Det}\, UA \to G(A) \ . \tag{6.7}$$

For $\mathcal{O}$ local we use Theorem 2 (iv) instead and take T as the map induced by ρ_s. A similar description to the local one then also applies to $\underline{P}_A \circ T$ in the global case.

For the statement of the next theorem recall the definition of the Hermitian form m_A^n on A^n, given in example (b) near the end of §5, with

$$m_A^n \left((a_1,\dots,a_n),\ (b_1,\dots,b_n)\right) = \Sigma\, \bar{a}_i\, b_i \ .$$

As already in that example we view A^n as canonically embedded in A^n.

If H is a subgroup of a group $\mathrm{Hom}_{\Omega_F}(K_{A,F}, \cdot)$ we shall write $\mathrm{Ker}\, \rho_s|H$ for the kernel of the restriction of ρ_s to H.

Theorem 4. (i) _If_ M _is a l.f.p. torsion_ A-_module, say_ $M = A^n/X$ _with_ $X \subset A^n$ _then_

$$T([M]) = d((X, m_A^n)) \ .$$

(ii) _For_ $\mathcal{O}$ _global we have a commutative diagram with exact columns_

$$\begin{array}{ccc}
1 & & 1 \\
\downarrow & & \downarrow \\
\operatorname{Ker} \rho_s|(\operatorname{Det} UA \cap \operatorname{Det} A^*) & \longrightarrow & \operatorname{Ker} \rho_s|\operatorname{Det} UA \\
\downarrow & & \downarrow \\
\operatorname{Ker} \rho_s|\operatorname{Det} A^* & \longrightarrow & \operatorname{Ker} \rho_s|\operatorname{Det} JA \\
\downarrow & & \downarrow \\
K_oT(A) & \longrightarrow & K_oT(A) \\
T\downarrow & & \underline{P}T\downarrow \\
HCl(A) & \xrightarrow{\underline{P}_A} & \operatorname{Ad} HCl(A) \\
\Phi\downarrow & & \operatorname{Ad}\Phi\downarrow \\
HCl(A) & \xrightarrow{\underline{P}_A} & \operatorname{Ad} HCl(A) \\
\downarrow & & \downarrow \\
1 & & 1
\end{array}$$

<u>and for</u> o <u>local an exact sequence</u>

$$1 \to \operatorname{Ker} \rho_s|\operatorname{Det} A^* \to \operatorname{Ker} \rho_s|\operatorname{Det} A^* \to K_oT(A) \xrightarrow{T} HCl(A) \xrightarrow{\Phi} HCl(A) \to 1 .$$

<u>Remark 1.</u> For o local

$$\operatorname{Ker} \rho_s|\operatorname{Det} A^* = \operatorname{Ker} \rho_s|\operatorname{Hom}_{\Omega_F}(K_{A,F}, F_c^*) .$$

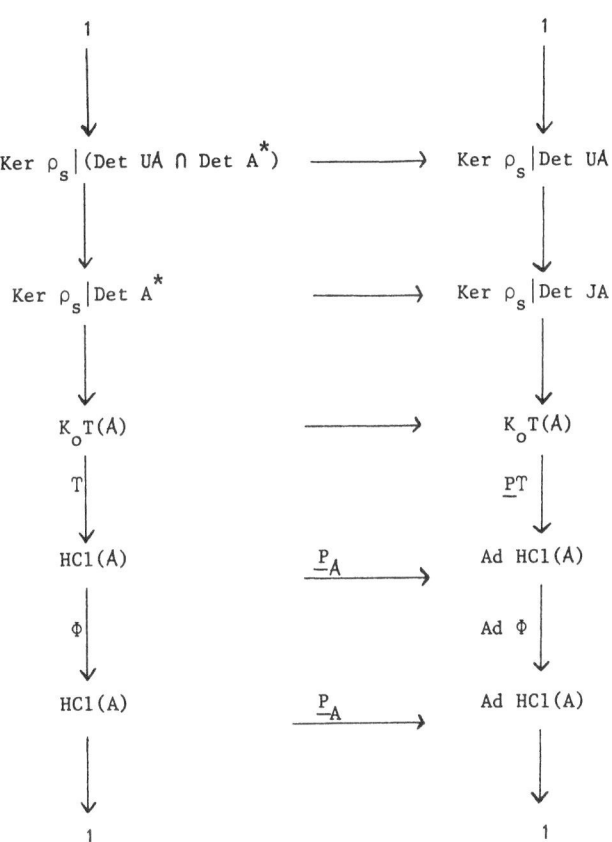

Remark 2: There is a more natural choice, to replace above the groups $\operatorname{Ker} \rho_s | H$. We define the group of automorphisms of the involution algebra $A, ^-$ by

$$\operatorname{Aut}(A, ^-) = [a \in A^* \mid a\bar{a} = 1] , \tag{6.8}$$

and define similarly $\operatorname{Aut}(\mathcal{A}, ^-), \operatorname{Aut}(JA, ^-)$.

Then we have

Supplement to Theorem 4. $\operatorname{Det} \operatorname{Aut}(A, ^-) \subset \operatorname{Ker} \rho_s | \operatorname{Det} A^*$, (and analogously for $\mathcal{A}$ and JA) .

Indeed, if (T, j) defines a symplectic representation, then for $a \in \operatorname{Aut}(A, ^-)$

$$1 = T(a)\, T(\bar{a}) = T(a)\, T(a)^j ,$$

whence by Proposition 3.1, $\operatorname{Det} T(a) = 1$, $\operatorname{Det}(a) \in \operatorname{Ker} \rho_s$.

The question is then whether e.g., we can replace in Theorem 4, $\operatorname{Ker} \rho_s | \operatorname{Det} A^*$ by $\operatorname{Det}(\operatorname{Aut}(A, ^-)$, i.e., whether

$$\operatorname{Ker} \rho_s | \operatorname{Det} A^* = \operatorname{Det} \operatorname{Aut}(A, ^-) \ ? \tag{6.9}$$

We shall return to this in III. In "one" case this will be seen to be false.

Proof of Theorem 4. We shall prove (i) for $\mathcal{O}$ global. The local case only differs in the notation. Let for each p, $X_p = a_p \mathcal{A}_p^n$, $a_p \in GL_n(\mathcal{A}_p)$. Then in $\operatorname{Det} JA/\operatorname{Det} U\mathcal{A}$, $[M]$ is represented by $\operatorname{Det}(a)$. Moreover by the example (a) at the end of §5, and as the discriminant matrix $(h(u_i, u_j))$ of $m_{\mathcal{A}}^n$ is 1 , we get $\operatorname{Pf}(h(u_i, u_j)) = 1$. Thus, in $G(\mathcal{A})$, $d((X, m_{\mathcal{A}}^n))$ is represented by $(\operatorname{Det}(a), 1)$, which is exactly what we had to prove.

(ii) The surjectivity of Φ is obvious in both the local and the global case (cf. (6.2, (6.3)). We now first proceed with the case $\mathcal{O}$ global. Exactness of the left sequence at $\mathrm{HCl}(A)$ is again quite clear. Now let $f \in \mathrm{Det}\, JA$ represent an element of $\mathrm{Ker}\, T$. This is equivalent with saying that $\exists\, g \in \mathrm{Det}\, A^*$ with $g^s = 1$, $g^{-1}f \in \mathrm{Det}\, UA$. In other words it means that $f \bmod \mathrm{Det}\, UA = g \bmod \mathrm{Det}\, UA$, $g \in \mathrm{Ker}\, \rho_s | \mathrm{Det}\, A^*$. Thus we have exactness of the left sequence at $K_o TA$. The remaining assertions for $\mathcal{O}$ global are easily checked, the exactness of the right sequence also following from the local result.

For $\mathcal{O}$ local the exactness is almost immediate by applying the snake lemma to the commutative diagram with exact rows and columns

$$\begin{array}{ccccccccc}
1 & \to & \mathrm{Det}(A^*) & \to & \mathrm{Hom}_{\Omega_F}(K_{A,F}, F_c^*) & \to & K_oT(A) & \to & 1 \\
 & & \downarrow & & \downarrow{\scriptstyle \rho_s} & & \downarrow & & \\
1 & \to & \mathrm{Det}^s A^* & \to & \mathrm{Hom}_{\Omega_F}(K^s_{A,F}, F_c^*) & \to & \mathrm{HCl}(A) & \to & 1 \\
 & & \downarrow & & & & & & \\
 & & 1 & & & & & &
\end{array}$$

.

There is also a homomorphism from $\mathrm{HCl}(A)$ to $K_oT(A)$, which generalises the map from $\mathrm{HCl}(\mathcal{O})$ to $K_oT(\mathcal{O})$, the latter being the group of fractional ideals of $\mathcal{O}$. In terms of the classical situation ($A = \mathcal{O}$, trivial involution) this map takes the "strong" discriminant as defined e.g., in [F1], i.e., modulo the unit squares ($\mathcal{O}$ local), or unit idele squares ($\mathcal{O}$ global), into the discriminant ideal.

Let (X,h) be a Hermitian A-module, and denote by $\hat{X}_h$ the dual of X under h, i.e.,

$$\hat{X}_h = [u \in X \otimes_A A \mid h(X,u) \subset A] \quad . \tag{6.10}$$

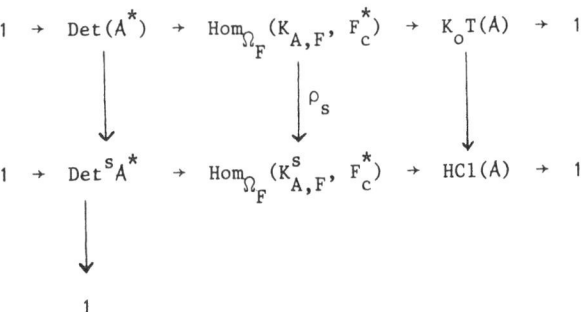

We define a class $[\hat{X}_h/X] \in K_oT(A)$. If actually $\hat{X}_h \supset X$ then indeed $\hat{X}_h/X$ is a l.f.p.torsion module. Otherwise take $[\hat{X}_h/X] = [\hat{X}_h/Y] - [X/Y]$, where Y is locally free, spans $X \otimes_A A$ and $Y \subset X \cap \hat{X}_h$. Clearly, the map $(X,h) \longmapsto [\hat{X}_h/X]$ yields a homomorphism

$$d' \colon K_oH(A) \to K_oT(A)\ . \tag{6.11}$$

Next recall the definition of the map

$$\mathrm{Nr}\colon \mathrm{Hom}_{\Omega_F}(K^s_{A,F}, F_c^*) \to \mathrm{Hom}_{\Omega_F}(K_{A,F}, F_c^*)\ ,$$

given in (4.6) (4.7). Supposing $\mathfrak{o}$ is local, we identify as usual $\mathrm{HCl}(A)$ as a quotient of the domain of Nr (via Proposition 5.1), and $K_oT(A)$ as a quotient of the range of Nr.

6.1 <u>Proposition:</u> <u>Suppose</u> $\mathfrak{o}$ <u>is local. The map</u> Nr <u>induces the homomorphism</u> ν <u>in the commutative diagram</u>

$$\begin{array}{ccc} & K_oH(A) & \\ {\scriptstyle d}\swarrow & & \searrow{\scriptstyle d'} \\ \mathrm{HCl}(A) & \xrightarrow{\ \nu\ } & K_oT(A) \end{array}$$

<u>Proof:</u> We consider the tentative commutative diagram with exact rows

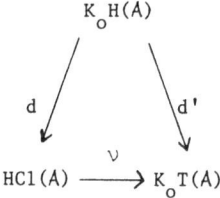

$$\begin{array}{ccccccccc}
1 & \to & \mathrm{Det}^s A^* & \to & \mathrm{Hom}_{\Omega_F}(K^s_{A,F}, F_c^*) & \to & \mathrm{HCl}(A) & \to & 1 \\
 & & \downarrow & & \downarrow \mathrm{Nr} & & \downarrow \nu & & \\
1 & \to & \mathrm{Det}\, A^* & \to & \mathrm{Hom}_{\Omega_F}(K_{A,F}, F_c^*) & \to & K_o T(A) & \to & 1
\end{array} \qquad (6.12)$$

and in order to show that there is a map ν making the right hand square commute we have to prove that Nr maps $\mathrm{Det}^s A^*$ into $\mathrm{Det}\, A^*$. But if $f \in \mathrm{Det}^s A^*$, say $f = \mathrm{Det}^s(a)$ with $a \in A^*$ then indeed by Proposition 1.1, we have $\mathrm{Nr}\, f = \mathrm{Det}\,(a\,\bar{a})$, and here $a\,\bar{a} \in A^*$, as $A = \bar{A}$.

Next let $\{x_i\}$ be a free basis of X over A and let $\{y_i\}$ be the dual basis of $\hat{X}_h$, i.e., we have

$$h(x_i, y_j) = \delta_{ij} \qquad \text{(Kronecker delta)}.$$

If we now write

$$x_k = \sum_j y_j\, a_{jk}, \qquad a_{jk} \in A,$$

then

$$h(x_i, x_k) = \sum_j h(x_i, y_j)\, a_{jk} = a_{ik}.$$

Thus $[\hat{X}_h/X]$ in $K_o T(A)$ is represented by $\mathrm{Det}(h(x_i, x_j))$, while $d((X,h))$ is represented by $\mathrm{Pf}(h(x_i, x_j))$. The commutativity of the diagram in 6.1 is now a consequence of Proposition 4.6.

We now consider the action of the group of order 2 via the automorphism $^-$ induced on various Hom groups and classgroup by the involution. We shall write H^i for the cohomology with respect to this action and $\hat{H}^o$ for the zero dimensional Tate cohomology. The involution will not be indicated explicitly in our notation, unless there is danger of confusion.

The kernel and cokernel of ν are best discussed in the context of indecomposable involution algebras. We shall however mention here some results. Recall the definitions of Nr and Tr (see (4.6), (4.7)).

6.2 *Proposition: Suppose* $\mathcal{O}$ *is local*

(i) Im Nr *and* Im ν , *as well as all groups in the top row of* (6.12) *are elementwise fixed under* $^{-}$.

(ii) *There is an exact sequence*

$$\mathrm{Hom}_{\Omega_F}(K^s_{A,F}/\mathrm{Tr}(K_{A,F}), \pm 1)$$

$$\to \mathrm{Ker}\ \nu \to \hat{H}^o(\mathrm{Det}\ A^*) \to \mathrm{Cok}\ H^o(\mathrm{Nr}) \to \mathrm{Cok}\ H^o(\nu) \to H^1(\mathrm{Det}\ A^*) \ .$$

(iii) *If* $K^s_{A,F} = \mathrm{Tr}(K_{A,F})$, *then*

$\mathrm{Ker}\ \nu \cong \hat{H}^o(\mathrm{Det}\ A^*)$, *and* $\mathrm{Cok}\ H^o(\nu)$ *embeds in* $H^1(\mathrm{Det}\ A^*)$.

Proof: (i) is obvious. Accordingly we now get from (6.12) a commutative diagram

$$\begin{array}{ccccccccc}
1 \longrightarrow & \mathrm{Det}^s A^* & \longrightarrow & \mathrm{Hom}_{\Omega_F}(K^s_{A,F}, F^*_c) & \longrightarrow & \mathrm{HCl}(A) & \longrightarrow & 1 \\
& \downarrow & & \downarrow{\scriptstyle H^o(\mathrm{Nr})} & & \downarrow{\scriptstyle H^o(\nu)} & & \downarrow \\
1 \to & H^o(\mathrm{Det}\ A^*) & \to & H^o(\mathrm{Hom}_{\Omega_F}(K_{A,F}, F^*_c)) & \to & H^o(K_oT(A)) & \to & H^1(\mathrm{Det}\ A^*)
\end{array}$$

with exact rows. To get (ii) we shall apply the snake Lemma. As $\mathrm{Det}(a\,\bar{a}) = \mathrm{Det}(a)\,\overline{\mathrm{Det}(a)}$ the cokernel of the left hand column is $\hat{H}^{o}(\mathrm{Det}\,A^{*})$. It remains to be shown that

$$\mathrm{Hom}_{\Omega_F}(K^{s}_{A,F}/\mathrm{Tr}(K_{A,F}),\ \pm 1) \cong \mathrm{Ker\ Nr}. \tag{6.13}$$

Indeed Ker Nr consists of the maps g with $g\,(\mathrm{Tr}\,\chi) = 1$, i.e., effectively maps $g\colon K^{s}_{A,F}/\mathrm{Tr}(K_{A,F}) \to F^{*}_{c}$. But $K^{s}_{A,F}/\mathrm{Tr}(K_{A,F})$ is killed by 2, and therefore $\mathrm{Im}\,g \subset \pm 1$. Hence the result.

Assertion (iii) now follows from (ii) and by observing moreover that, under the stated hypothesis, $H^{o}(\mathrm{Nr})$ is an isomorphism.

Remark 1: The isomorphism for $\mathrm{Ker}\,\nu$ under (iii) generalises the result for $A = \mathfrak{o}$ and the trivial involution. In this latter case the kernel of the transition from the "strong" discriminant to the discriminant ideal is the group of units modulo unit squares.

Remark 2: The appearance of the group $\mathrm{Hom}_{\Omega_F}(K^{s}_{A,F}/\mathrm{Tr}(K_{A,F}),\ \pm 1)$ is rather significant. In the arithmetic theory of [F7] and of [F8] this group plays an important role. In particular it houses the symplectic root numbers.

Now we return to the global case. We suppose $\mathfrak{o}$ to be global and denote by ν_p the map defined above for the local order A_p, p being a finite prime divisor of F. For almost all p, the group $\mathrm{Hom}_{\Omega_{F_p}}(K_{A_p,F_p},\ Y(F_{p,c}))$ coincides with $\mathrm{Det}\,A^{*}_{p}$ - this is certainly the case for all p for which A_p is a sum of matrix rings over fields and A_p is a maximal order. Therefore for almost all p the group $\mathrm{HCl}''(A_p)$ (see (5.10)) lies in $\mathrm{Ker}\,\nu_p$. Hence the ν_p for all finite p, together with the null maps $\mathrm{HCl}(A_p) \to 1$ for infinite p, yield a global homomorphism

$$\nu\colon \mathrm{Ad}\ \mathrm{HCl}(A) \to K_oT(A)\ . \tag{6.14a}$$

(Use 5.2 and Theorem 1 (i)).

Corollary to Proposition 6.1 The diagram

$$\begin{array}{ccc} & K_o\,H(A) & \\ {\scriptstyle \underline{P}d}\swarrow & & \searrow{\scriptstyle d'} \\ \mathrm{Ad}\ \mathrm{HCl}(A) & \xrightarrow{\ \nu\ } & K_o\,T(A) \end{array}$$

(o global) commutes.

We also have a map

$$\nu_p\colon \mathrm{HCl}(A_p) = \mathrm{HCl}(A_p) \to K_oT(A_p) = K_oT(A_p)$$

for infinite p - for the right hand side see I (3.1). It is again induced by the map Nr . Taking now the ν_p , for all p , as local components we get a global homomorphism

$$\tilde{\nu} : \mathrm{Ad}\ \mathrm{HCl}(A) \to \widetilde{K_oT(A)}\ , \tag{6.14b}$$

with $\widetilde{K_oT(A)}$ defined in I. (3.2), (3.5). The map ν of (6.14a) is then obtained by composing $\tilde{\nu}$ with the projection $\widetilde{K_oT(A)} = K_oT(A) \times K_oT_\infty(A) \to K_oT(A)$ (cf. I. (3.5)). Alternatively $\tilde{\nu}$ can be viewed as the homomorphism

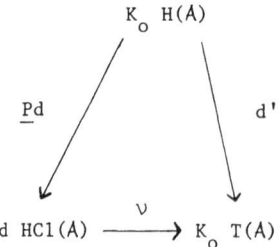

$$\frac{\mathrm{Hom}_{\Omega_F}(K^s_{A,F}, J(F_c))}{\mathrm{Det}^s \ U\mathring{A}} \longrightarrow \frac{\mathrm{Hom}_{\Omega_F}(K_{A,F}, J(F_c))}{\mathrm{Det} \ U\mathring{A}}$$

induced by the map Nr . See here I (3.6).

One can of course also derive a global version of Proposition 6.2.

From now on, and until the end of this §6, o is assumed to be global. In I (1.3) we defined a map $\delta: K_oT(A) \to K_o(A)$, which gives rise to a map $K_oT(A) \to Cl(A)$, again to be denoted by δ. Next one verifies that the map $(X,h) \longmapsto (X)$ yields a homomorphism

$$\delta': K_oH(A) \to Cl(A) . \tag{6.15}$$

Finally composing the component projection

$$G(A) \to \mathrm{Det} \ J\mathring{A}/\mathrm{Det} \ U\mathring{A}$$

with

$$\mathrm{Det} \ J\mathring{A}/\mathrm{Det} \ U\mathring{A} \to Cl(A) ,$$

induces a homomorphism

$$\delta'' : HCl(A) \to Cl(A) \tag{6.16}$$

6.3 Proposition: Let o be global. (i) The diagram

$$\begin{array}{ccccc} K_o\, T(A) & \xrightarrow{T} & \mathrm{HCl}(A) & \xleftarrow{d} & K_o\, H(A) \\ & \searrow^{\delta} & \downarrow^{\delta''} & \swarrow_{\delta'} & \\ & & \mathrm{Cl}(A) & & \end{array}$$

commutes.

(ii) _The component injection_

$$\mathrm{Hom}_{\Omega_F}(K^s_{A,F},\ F^*_c) \to G(A)$$

yields exact sequences

$$(\mathrm{Det}\ A^* \cap \mathrm{Det}\ UA) \xrightarrow{\rho_s} \mathrm{Hom}_{\Omega_F}(K^s_{A,F},\ F^*_c) \to \mathrm{HCl}(A) \to \mathrm{Cl}(A) \to 1$$

$$(\mathrm{Det}\ A^* \cap \mathrm{Det}\ UA) \xrightarrow{\rho_s} \mathrm{Det}^s\ A^* \longrightarrow \mathrm{HCl}(A) \to \mathrm{Cl}(A) \times \mathrm{HCl}(A) \to 1\ .$$

Remark: We gave the last sequence mainly to indicate that the map $\mathrm{HCl}(A) \to \mathrm{Cl}(A) \times \mathrm{HCl}(A)$ which is really induced by the identity map on $G(A)$ does in general have a non-trivial kernel.

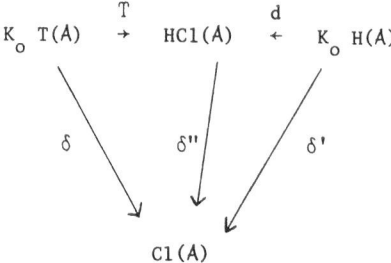

Proof of 6.3: (i) Let (X,h) be a Hermitian module, q the rank of X. Let $\{v_i\}$ be a basis of $V = X \otimes_A A$ over A, and let U be the free A-module on $\{v_i\}$. Then $d((X,h))$ is represented by $(\mathrm{Det}(a), \mathrm{Pf}(h(v_j,v_j)))$ where $X = aU$, $a \in \mathrm{Aut}_A(V)$. The image under δ'' is represented by $\mathrm{Det}(a)$, and under the isomorphism (vi) of Theorem 2, $\mathrm{Det}(a)$ also represents $(X) = [X] - [A^q] = [X] - [U]$. So the right hand triangle commutes. For the left hand triangle, let $M = U/X$ with $U = A^q$ - this is of course always possible for appropriate q. Then use Theorem 4. (i) to go via $K_oH(A)$ as above - choosing the given basis of U as a basis of $V = X \otimes_A A$, with $h = m_A^n$. The commutativity is then immediate from the definition of δ.

(ii) The map $HCl(A) \to Cl(A) \times HCl(A)$ is induced by the identity map on $G(A)$ and must therefore be surjective. Hence, or by (i) and the surjectivity of δ or of δ', also δ'' is surjective. Next (f mod Det UA, g) $\in G(A)$ will represent an element in $\operatorname{Ker}\delta''$ precisely if $f \in \operatorname{Det} A^* \operatorname{Det} UA$. But then the given class is also represented by an element (1 mod Det UA, g_1) $\in G(A)$. This implies the exactness of the top sequence at $HCl(A)$.

Next $g \in \operatorname{Hom}_{\Omega_F}(K^s_{A,F}, F^*_c)$ will lie in the kernel of the map to $HCl(A)$, precisely if there is an element $f \in \operatorname{Det} A^*$, with $f^s = g$ and $f \in \operatorname{Det} UA$. This observation now gives the exactness of the top sequence. For the exactness of the bottom sequence we only have to verify that $\operatorname{Det}^s A^*$ is the kernel of the composite map

$$\operatorname{Hom}_{\Omega_F}(K^s_{A,F}, F^*_c) \to HCl(A) \to HCl(A) .$$

This however is obvious.

By composing the maps T, ν and $\underline{P}$ we obtain endomorphisms of $K_oT(A)$ and of $HCl(A)$ and $\operatorname{Ad} HCl(A)$. These correspond to the maps $f \mapsto f\bar{f}$ and $f \mapsto f^2$ on $\operatorname{Hom}_{\Omega_F}(K_{A,F}, \cdot)$ and on $\operatorname{Hom}_{\Omega_F}(K^s_{A,F}, \cdot)$ respectively. The details are left to the reader. We have also not discussed the extension of T to a homomorphism $K_o\, T_\infty(A) \to HCl(A)$ and by composition to $K_o\, T_\infty(A) = K_o\, T_\infty(A) \to HCl(A)$.

§7 Pulling back discriminants

We shall consider two related problems, namely the Hasse principle for discriminants and the surjectivity of the discriminant map.

Note first

7.1 <u>Proposition:</u> <u>The diagrams</u>

$$\begin{array}{ccc} K_0H(A) & \to & HCl(A) \\ \downarrow & & \downarrow \\ K_0H(\mathrm{A}) & \to & HCl(\mathrm{A}) \end{array}\quad , \quad \begin{array}{ccc} K_0H(A) & \to & HCl(A) \\ \downarrow & & \downarrow \\ K_0H(A_p) & \to & HCl(A_p) \end{array}\quad , \quad \begin{array}{ccc} K_0H(\mathrm{A}) & \to & HCl(\mathrm{A}) \\ \downarrow & & \downarrow \\ K_0H(\mathrm{A}_p) & \to & HCl(\mathrm{A}_p) \end{array}$$

<u>commute</u> (o <u>global for the second and third diagram</u>).

This is obvious. By Propositions 5.2, 5.4, corresponding statements also follow for Ad $HCl(A)$, Ad $HCl(\mathrm{A})$ in place of $HCl(A)$, $HCl(\mathrm{A})$.

<u>Terminology:</u> Suppose we have a homomorphism

$$h: HCl(A) \to \prod_j HCl(\mathrm{A}_j)$$

coming from changes $\mathrm{A} \to \mathrm{A}_j$ of orders, of the type we have already discussed. We shall say that discriminants can be <u>pulled back via</u> h if the following condition is satisfied: Whenever, for all j , $h(x)_j \in \mathrm{Im}\, d_{\mathrm{A}_j}$, where the subscript indicates the order for the definition of the discriminant, then $x \in \mathrm{Im}\, d_{\mathrm{A}}$.

<u>Theorem 5:</u> <u>Discriminants can be pulled back via the following maps.</u>

(i) $HCl(A) \to HCl(\mathrm{A})$

(ii) <u>For global</u> o, $HCl(A) \to \prod_p HCl(A_p)$ (<u>product over all</u> p)

(iii) <u>For global</u> o, $HCl(\mathrm{A}) \to \prod_{p|\infty} HCl(\mathrm{A}_p)$ (<u>product over all infinite</u> p)

(iv) <u>Also,</u> d <u>is surjective onto</u> $HCl(\mathrm{A})$ <u>when</u> o <u>is local</u>

(i.e., *when* F *is a finite extension of* $\mathbb{Q}_p$, p *finite*).

Remark 1: In view of (iv) the product in (iii) may be taken over all p. (ii), (iii) and (iv) imply the *Hasse principle for discriminants*.

Remark 2: In view of Propositions 5.2 and 5.4 the right hand sides in (ii), (iii) may be replaced by the appropriate Ad HCl-groups.

We shall give here a proof of (i) and, under the hypothesis that (iii) and (iv) hold, also of (ii). Clearly for (iii) and (iv) one may assume that $(A, \bar{\ })$ is indecomposable as involution algebra, and under this hypothesis these assertions will be established in III §3, where we shall also give some more results.

Consider an element $x \in \mathrm{HCl}(A)$ with its image $\Phi x \in \mathrm{HCl}(A)$ being a discriminant, i.e., $\Phi x \in \mathrm{Im}\, d_A$. By Propositions 2.1 and 7.1, $\exists\, y \in \mathrm{Im}\, d_A$ with $\Phi y = \Phi x$. Therefore $x = yz$, $z \in \mathrm{Ker}\, \Phi$. By Theorem 4 (ii), $z \in \mathrm{Im}\, T$ and by Theorem 4 (i), $z \in \mathrm{Im}\, d_A$. Thus $x \in \mathrm{Im}\, d_A$. We have now established (i).

For (ii) observe first that the diagram

$$\begin{array}{ccc} \mathrm{HCl}(A) & \longrightarrow & \prod_p \mathrm{HCl}(A_p) \\ \Big\downarrow \Phi & & \Big\downarrow \prod_p \Phi_p \\ \mathrm{HCl}(A) & \longrightarrow & \prod_p \mathrm{HCl}(A_p) \end{array} \qquad (7.1)$$

commutes. Indeed let $(f,g) \in \mathrm{Det}\, JA \times \mathrm{Hom}_{\Omega_F}(K^s_{A,F}, F^*_c)$ represent a given element of $\mathrm{HCl}(A)$. Going round to $\prod_p \mathrm{HCl}(A_p)$ anticlockwise we get for given p an element represented by g_p, going round the other way one represented by $g_p f^s_p$. But $f^s_p \in \mathrm{Det}^s A^*_p$, and so cancels.

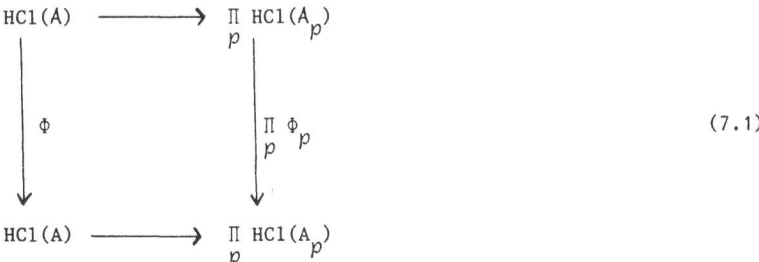

By (iii) we can pull back via the bottom row, and by (i) via the left hand column. Trivial diagram chasing now gives (ii).

§8 Unimodular modules

We shall briefly return to the situation considered in standard unitary K-theory, which is based on a generalisation of unimodular lattices. The contents of this section is thus in particular related to the relevant parts of C.T.C Wall's series of papers [Wa 5]. With A, $\mathcal{A}$ as before, we shall call a Hermitian $\mathcal{A}$-module (X,h) unimodular if h actually gives rise to a non-singular Hermitian form $X \times X \to \mathcal{A}$, or equivalently if $X = \hat{X}_h$ (the latter defined in (6.10)). Our purpose here is solely to indicate how the notion of unimodularity fits into our general theory.

The classes of unimodular $\mathcal{A}$-modules form a subgroup of $K_oH(\mathcal{A})$, which we shall denote by $K_o\mathcal{U}(\mathcal{A})$, and which is an image of the Grothendieck group of non-singular Hermitian forms over $\mathcal{A}$.

We define

$$\mathcal{U}HCl(A) = \text{Ker} \quad [HCl(A) \to K_oT(A)] . \tag{8.1}$$

If $\mathfrak{o}$ is local the map on the right is ν, as defined in Proposition 6.1, i.e., $\mathcal{U}HCl(A) = \text{Ker}\ \nu$. If $\mathfrak{o}$ is global then $HCl(A) \to K_oT(A)$ is the compositum of ν, as in (6.14a) and of $\underline{P}_A$.

8.1 Proposition: The discriminants of unimodular modules take values in $\mathcal{U}HCl(A)$, i.e., d maps $K_o\,\mathcal{U}(\mathcal{A})$ into $\mathcal{U}HCl(A)$.

Proof: Clearly $K_o\mathcal{U}(\mathcal{A}) \subset \text{Ker}[K_oH(\mathcal{A}) \to K_oT(A)]$ (see (6.11) for this map). For $\mathfrak{o}$ local now use Proposition 6.1 and similarly for $\mathfrak{o}$ global its Corollary.

Question 1: When is $K_o U(A) = \mathrm{Ker}[K_o H(A) \to K_o T(A)]$?

Question 2: Clearly

$$d(K_o H(A)) \cap U\mathrm{HCl}(A) \supset d(K_o U(A)) .$$

Is there equality?

Next we shall consider the effect of the map $\delta'' : \mathrm{HCl}(A) \to \mathrm{Cl}(A)$ on $U\mathrm{HCl}(A)$ - assuming from now on $\mathcal{O}$ to be global. For this purpose we define some further maps and groups. If G is a subgroup of $\mathrm{Hom}_{\Omega_F}(K_{A,F}, J(F_c))$ which admits the operator $^-$ induced by the involution, we shall write again $H^o(G)$ for the fixed point of $^-$ in G. The map $x \longmapsto x\bar{x}$ is a homomorphism $G \to H^o(G)$ and in fact

$$x\bar{x} = \mathrm{Nr}\, \rho^s(x) .$$

We shall now apply this to the relevant groups and we get a homomorphism

$$N\colon \mathrm{Cl}(A) = \mathrm{Det}\, JA/\mathrm{Det}\, UA.\mathrm{Det}\, A^* \to \tag{8.2}$$

$$H^o(\mathrm{Hom}_{\Omega_F}(K_{A,F}, J(F_c)))/H^o(\mathrm{Det}\, UA).H^o(\mathrm{Hom}_{\Omega_F}(K_{A,F}, F_c^*))H^o(\mathrm{Hom}_{\Omega_F}(K_{A,F}, J_\infty(F_c))) .$$

Here N is induced by the map $x \mapsto x\bar{x}$ of $\mathrm{Det}\ JA$ into $H^o(\mathrm{Hom}_{\Omega_F}(K_{A,F}, J(F_c)))$ and for any number field E, $J_\infty(E) = \prod_{p|\infty} E_p^*$ (product over the infinite prime divisors).

8.2 Proposition:

$$\delta'' \ (\mathcal{U}HCl(A)) \subset \text{Ker } N \ ,$$

and whenever $K^{s}_{A,F} = Tr(K_{A,F})$ then

$$\delta'' \ (\mathcal{U}HCl(A)) = \text{Ker } N \ .$$

(Tr was defined in (4.6)).

Remark: Suppose that

$$\text{Hom}_{\Omega_F}(K_{A,F}, J_\infty(F_c)) \subset \text{Det } \mathcal{U}A \ . \tag{8.3}$$

Then we have

$$\text{R.H.S. in (8.2)} = H^o(\text{Det } JA)/H^o(\text{Det } \mathcal{U}A) \ H^o(\text{Det } A^*) \ . \tag{8.4}$$

For, in this case also $\text{Hom}_{\Omega_F}(K_{A,F}, F_c^*) = \text{Det } A^*$. (8.3) will hold for instance if for no infinite prime $\mathfrak{p}$ of F , $A_{\mathfrak{p}}$ contains a matrix ring over $\mathbb{H}$.

Corollary 1. Suppose (8.3) to hold. If $c \in \delta''(\mathcal{U}HCl(A))$ then $c\bar{c} = 1$.

Corollary 2. Suppose (8.3) to hold. If the locally free A-module X admits a non-singular Hermitian form into A then $(X)(\bar{X}) = 1$.

Corollary 1 is immediate. For Corollary 2, we also use Proposition 8.1 and 6.3.

Example 1: Let $A = \mathcal{O}$, with trivial involution. The map N is $\mathrm{Cl}(\mathcal{O}) \overset{2}{\to} \mathrm{Cl}(\mathcal{O})$ and so $\mathrm{Ker}\, N = \mathrm{Cl}(\mathcal{O})_2$, the kernel on squaring.

Example 2: Let A/F be a quadratic field extension, with $\bar{\ }$ as the non trivial automorphism and let A be the integral closure of $\mathcal{O}$. Then $\mathrm{Ker}\, N = \mathrm{Ker}\, N_{A/F}$. ($N_{A/F}$ the norm).

In both these examples (8.3) holds, $K^s_{A,F} = \mathrm{Tr}(K_{A,F})$ and $d\colon K_0\mathcal{U}(A) \to \mathcal{U}\mathrm{HCl}(A)$ is surjective.

Proof of 8.2. Let

$$a \in \mathrm{Det}\, JA, \quad b \in \mathrm{Hom}_{\Omega_F}(K^s_{A,F}, F_c^*) . \tag{8.5}$$

The class of (a,b) in $\mathrm{HCl}(A)$ will lie in $\mathcal{U}\mathrm{HCl}(A)$ if, and only if, the projection of the element $\mathrm{Nr}((\rho_s\, a)b)$ of $\mathrm{Hom}_{\Omega_F}(K_{A,F}, J(F_c))$ onto the finite part $\mathrm{Hom}_{\Omega_F}(K_{A,F}, J_{fin}(F_c))$ lies in $\mathrm{Det}\, \mathcal{U}A$, where for a number field E , the finite idele group $J_{fin}(E)$ is $\mathrm{Ker}\,[J(E) \to J_\infty(E)]$. An equivalent statement is that

$$\mathrm{Nr}((\rho_s\, a)b) = uv, \quad u \in \mathrm{Det}\, \mathcal{U}A , \quad v \in \mathrm{Hom}_{\Omega_F}(K_{A,F}, J_\infty(F_c)). \tag{8.6}$$

Now the left hand side of (8.6) is fixed under $\bar{\ }$, hence so is the right hand side, and a moments thought shows that we may in fact assume that

$$u \in H^0(\mathrm{Det}\, \mathcal{U}A) , \quad v \in H^0(\mathrm{Hom}_{\Omega_F}(K_{A,F}, J_\infty(F_c))) . \tag{8.7}$$

The image of the class of (a,b) under δ'' is represented by a, and we now have $a\bar{a} = \mathrm{Nr}\, b^{-1} \cdot u \cdot v$, and this element lies in the denominator of the right hand side of (8.2). For the converse assume this is the case, i.e., that

$$a\bar{a} = u\, v\, w\,, \tag{8.8}$$

with u,v as above in (8.7), $w \in \mathrm{Hom}_{\Omega_F}(K_{A,F}, F_c^*)$. Then $\bar{w} = w$. Now suppose that $K_{A,F}^s = \mathrm{Tr}(K_{A,F})$. Then Tr defines an isomorphism $K_{A,F}/K_{A,F}^- \cong K_{A,F}^s$, where $K_{A,F}^-$ is the subgroup of $K_{A,F}$ of elements $\chi - \bar{\chi}$. Therefore Nr gives rise to an isomorphism

$$\mathrm{Hom}_{\Omega_F}(K_{A,F}^s, F_c^*) \cong H^0(\mathrm{Hom}_{\Omega_F}(K_{A,F}, F_c^*))$$

(see Proposition 6.2). Hence the element w in (8.8) is of form $w = \mathrm{Nr}\, b$, with b as in (8.5), and now we see that (a,b) represents a class in $\mathcal{U}\mathrm{HCl}(A)$, whose image is the given class in $\mathrm{Ker}\, N$, as represented by a.

§9. Products

For completeness sake we mention here the obvious results for products of orders and algebras. Suppose that the F-algebra

$$A = \prod_i A^{(i)} \tag{9.1}$$

is the product of F-algebras $A^{(i)}$, where of course the index set $\{i\}$ is finite, and that

$$A = \prod_i A^{(i)} \tag{9.2}$$

is correspondingly the product of orders $A^{(i)}$ in $A^{(i)}$. Then

$$K_o(A) = \prod_i K_o(A^{(i)}),\ Cl(A) = \prod_i Cl(A^{(i)}),\ K_oT(A) = \prod_i K_oT(A^{(i)}) . \tag{9.3}$$

If we moreover assume (9.1) to be a product of involution algebras, and the $A^{(i)}$ be invariant under the involutions then

$$K_oH(A) = \prod_i K_oH(A^{(i)}),\quad HCl(A) = \prod_i HCl(A^{(i)}) \tag{9.4}$$

and in the global case

$$\text{Ad HCl}(A) = \prod_i \text{Ad HCl}(A^{(i)}),\quad \text{Ad HCl}(A) = \prod_i \text{Ad HCl}(A^{(i)}) . \tag{9.5}$$

Finally all the maps in the preceding chapters between these groups preserve products. Analogous remarks apply of course to $K_{A,F}$, $K^s_{A,F}$ etc, and the associated Hom groups.

CHAPTER III. INDECOMPOSABLE INVOLUTION ALGEBRAS

An indecomposable involution algebra is one which cannot be written as a product of two involution algebras. The reason for looking at these is twofold. Firstly certain results are best stated in these terms and many proofs reduce partly or completely to this case, in particular of course all those which only involve the algebra, not the order. Secondly indecomposable involution algebras provide explicit illustrations for general result, indicating how these extend known classical ones, and allowing explicit computations. The study of the concrete situations arising here leads by necessity back from the language of Hom-groups to a more traditional one of groups of elements or ideles in the centre, e.g., replacing our determinants by reduced norms. This has the advantage that the reader can interpret the general theory in terms more familiar to him. It will become apparent that in this reinterpretation a variety of distinct cases have to be considered separately. This exhibits once more the power of our formalism of Hom groups, determinants and Pfaffians to provide a unified language for all these different situations. E.g. our discriminant generalises various invariants for particular indecomposable involution algebras, which occur in the literature.

Throughout this chapter it will again be assumed that we have one of our three basic cases: $\mathfrak{o}$ global, $\mathfrak{o}$ local, or $\mathfrak{o} = F$ a field - although clearly many results generalise.

§1. Dictionary

With $(A, \bar{\ })$ as before we assume here furthermore that it is an indecomposable involution algebra. Throughout $C = \mathrm{cent}(A)$, and then $(C, \bar{\ })$ is an indecomposable involution algebra. The sub-algebra

$H^o(C)$ fixed elementwise under $\bar{\ }$ is a field and we shall throughout this whole Chapter III assume that

$$H^o(C) = F . \tag{1.1}$$

By I. Proposition 2.1, this implies no loss of generality.

Assume first that $C = F$. Then A is a simple algebra, $G_{C,F}$ is a one element set. The irreducible representation of A, unique to within equivalence is given by a composed map

$$T_1 : A \to A \otimes_F F_c \cong M_m(F_c) , \tag{1.2}$$

where the first map is induced by the embedding $F \to F_c$. We shall denote by χ_1 its class. Then

$$K_{A,F} = \mathbb{Z}\ \chi_1 \tag{1.3}$$

and in view of the uniqueness of χ_1 we have $\chi_1 = \overline{\chi_1}$ and $\bar{\ }$ acts trivially on $K_{A,F}$. Moreover the involution $\bar{\ }$ extends uniquely to an F_c-involution $\bar{\ }$ of $A \otimes_F F_c$ and the isomorphism with $M_m(F_c)$ transfers this to an involution j of the matrix ring, unique to within inner automorphism. Thus T_1 yields a representation of A as an involution algebra. If j is orthogonal (symplectic) we call $(A,\bar{\ })$ itself <u>orthogonal</u> (<u>symplectic</u>). The distinction of the two cases is easy to check in terms of $(A,\bar{\ })$ itself. We have

$$\dim_F(H^o(A)) = \begin{cases} m(m+1)/2, & \text{orthogonal case,} \\ m(m-1)/2, & \text{symplectic case,} \end{cases}$$

where $\dim_F A = m^2$. (Our definition of the "orthogonal", "symplectic", or "unitary" (see below) property is equivalent to that of Wall (cf. [Wa 2])).

Next we have clearly

$$K^s_{A,F} = \begin{cases} 2K_{A,F} & \text{(orthogonal case)}, \\ K_{A,F} & \text{(symplectic case)}. \end{cases} \tag{1.4}$$

From (1.3), (1,4) we get

1.1 Proposition: (Orthogonal and symplectic case) *Let* G *be an* Ω_F*-module. There is a commutative diagram*

$$\begin{array}{ccc} \mathrm{Hom}_{\Omega_F}(K_{A,F}, G) & \cong & G^{\Omega_F} \\ \downarrow \rho_s & & \downarrow \ell \\ \mathrm{Hom}_{\Omega_F}(K^s_{A,F}, G) & \cong & G^{\Omega_F} \end{array},$$

where the top row, and in the symplectic case the bottom row as well, is given by $f \mapsto f(\chi_1)$ *and in the orthogonal case the bottom row is given by* $f \mapsto f(2\chi_1)$, *and where* ℓ *is the identity in the symplectic case and is* $x \mapsto x^2$ *(multiplicative notation) in the orthogonal case.*

We shall also use I. Proposition 2.7. to replace determinants again by reduced norms. Indeed the equation

$$\mathrm{Det}_{\chi_1}(a) = g_{\chi_1} \mathrm{nrd}(a) \tag{1.5}$$

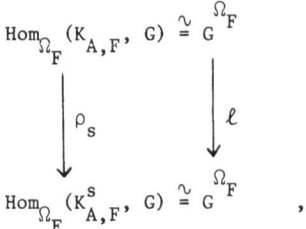

is really the definition of nrd(a)! The map g_{χ_1} is now just the embedding $F \to F_c$.

Note also for applications of the Proposition that if $G = H(F_c)$, H a functor satisfying the hypothesis of I Proposition 2.4, then $G^{\Omega_F} = H(F)$.

Next suppose that $F \neq C$. We shall then call $(A, {}^-)$ unitary. Here two sub cases arise. Either we have

$$C = F \times F, \quad A = B \times B^{op} , \tag{1.6}$$

where B is a simple F-algebra, B^{op} its opposite and the involution interchanges the two factors of C and of A . We call this the split case. Or else A is simple, and C is a quadratic extension of F , with $^-$ acting as the non-trivial automorphism of C/F . This is the non-split case. Note that in the latter case there is associated with C a subgroup of Ω_F of index 2, which fixes the unique image of C under any embedding of C into F_c over F . We denote this group by Ω_C . In both cases $C \otimes_F F_c \cong F_c \times F_c$ and hence $A \otimes_F F_c \cong M_m(F_c) \times M_m(F_c)$. Accordingly we get two non equivalent representations $T_1, \bar{T}_1$. We write $\chi_1, \overline{\chi_1}$ for these classes, where indeed $^-$ interchanges the two. Now we have

$$K_{A,F} = \mathbb{Z}\,\chi_1 \oplus \mathbb{Z}\,\bar{\chi}_1, \quad K^s_{A,F} = \mathbb{Z}\,(\chi_1 + \bar{\chi}_1) . \tag{1.7}$$

In the split case Ω_F acts trivially on $K_{A,F}$, in the non-split case Ω_F acts via the surjections on the group of order 2 generated by $^-$, with kernel Ω_C . Thus if G is an Ω_F-module we now have

1.2 Proposition: (Unitary case) There are commutative diagrams

$$\begin{array}{ccc}
\mathrm{Hom}_{\Omega_F}(K_{A,F}, G) & \cong & G^{\Omega_F} \times G^{\Omega_F} \\
\Big\downarrow \rho_s & & \Big\downarrow N_{C/F} \\
\mathrm{Hom}_{\Omega_F}(K^s_{A,F}, G) & \cong & G^{\Omega_F}
\end{array} \qquad \text{(split case)}$$

$$\begin{array}{ccc}
\mathrm{Hom}_{\Omega_F}(K_{A,F}, G) & \cong & G^{\Omega_C} \\
\Big\downarrow \rho_s & & \Big\downarrow N_{C/F} \\
\mathrm{Hom}_{\Omega_F}(K^s_{A,F}, G) & \cong & G^{\Omega_F}
\end{array} \qquad \text{(non split case)}$$

where the bottom rows are given by the evalutation map $f \mapsto f(\chi_1 + \bar{\chi}_1)$, the top row in the diagram for the split case by the evaluation map $f \mapsto (f(\chi_1), f(\bar{\chi}_1))$ and in the diagram for the non-split case by the evaluation map $f \mapsto f(\chi_1)$ for a given choice of χ_1. Moreover $N_{C/F}$ is the map $(x_1,x_2) \mapsto x_1 x_2$ (split case), $x \mapsto x\bar{x}$ (non-split case).

Remark 1: In the split case we may view G^{Ω_F} as embedded diagonally in $G^{\Omega_F} \times G^{\Omega_F}$. Then we can also write $N_{C/F}(x_1,x_2) = (x_1 x_2, x_1 x_2) =$ $= (x_1,x_2)(x_2,x_1) = (x_1,x_2)\overline{(x_1,x_2)}$, as in the non-split case.

Remark 2: The formulae get neater if $G^{\Omega_E} = H(E)$, for extensions E of F, with H a functor satisfying the hypothesis of I. Proposition 2.4. For, then we get in the split case, with the usual convention, that $G^{\Omega_F} \times G^{\Omega_F} = H(C)$, i.e., both right hand columns in 1.2. now represent $N_{C/F}: H(C) \to H(F)$.

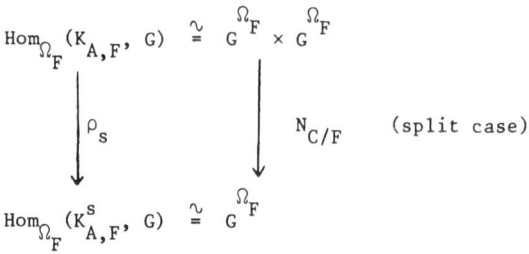

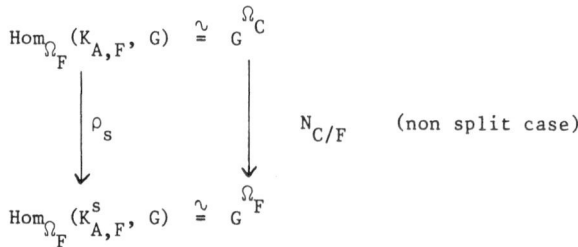

The previous comment on determinants and reduced norms (taken componentwise in the split case) applies again. If A splits we get

$$(\mathrm{Det}_{\chi_1}(a),\quad \mathrm{Det}_{\bar{\chi}_1}(a)) = (g_{\chi_1}\,\mathrm{nrd}(a),\quad g_{\bar{\chi}_1}\,\mathrm{nrd}(a))\ , \tag{1.8}$$

while in the non-split case (1.5) will still hold.

As an illustration, and also to answer a question which arises naturally we now return to the groups $\mathrm{HCl}''(\mathfrak{A}_p)$ defined in II, (5.10). For the moment we consider again a not necessarily indecomposable involution algebra $(A,\bar{\ })$ and a $\bar{\ }$ invariant order $\mathfrak{A}$ with o global.

By II. Proposition 5.2. (ii), we have an injective homomorphism

$$\coprod \mathrm{HCl}(\mathfrak{A}_p) \to \mathrm{Ad\ HCl}(\mathfrak{A}) \tag{1.9}$$

and we ask: when is this an isomorphism?

1.3 <u>Proposition</u>: (1.9) <u>is an isomorphism if and only if none of the indecomposable involution components of</u> $(A,\bar{\ })$ <u>are orthogonal</u>.

<u>Proof</u>: Observe that (1.9) is an isomorphism precisely when all but a finite number of the $\mathrm{HCl}''(\mathfrak{A}_p)$ vanish. As $\mathfrak{A}_p$ is a maximal order for all but a finite number of p, we may in fact suppose that $\mathfrak{A}$ is a maximal order, thus the product of maximal orders in indecomposable components. Hence we may also assume that $(A,\bar{\ })$ is in fact indecomposable (and from now on this will again be assumed). Then for all finite p, $(A_p,\bar{\ })$ is indecomposable. As $\mathfrak{A}_p$ is maximal, we have $\mathrm{nrd}\,\mathfrak{A}_p^* = \mathfrak{C}_p^*$, where $\mathfrak{C}_p$ is the maximal order of C_p. Hence, using our dictionary (and extending it somewhat)

$$
\mathrm{HCl}''(A_p) = \begin{cases} o_p^{*}/o_p^{*2} & \text{(orthogonal case)}, \\ 1 & \text{(symplectic case)}, \\ o_p^{*}/\mathrm{N}_{C/F}\, C_p^{*} & \text{(unitary case)}. \end{cases}
$$

In the symplectic case the group thus always vanishes; in the unitary case it also vanishes except when C is a field and p is genuinely ramified in C , and this can happen at only finitely many primes p . On the other hand in the orthogonal case $\mathrm{HCl}''(A_p)$ is always at least of order 2 .

§2. The map $\underline{P}$

We shall describe the kernels and cokernels of the maps $\underline{P}_{A}$ and $\underline{P}_{\mathrm{A}}$ in the indecomposable situation. For $\underline{P}_{\mathrm{A}}$ everything of course reduces to the indecomposable case, and by II. Proposition 5.4, the same is true for Cok $\underline{P}_{A}$.

We have to introduce some additional notation. Throughout o is global. Then

$$
\mathcal{J}(F) = J(F)/F^{*} \tag{2.1}
$$

is the idele classgroup of F , and we put

$$
\mathcal{J}(F \times F) = \mathcal{J}(F) \times \mathcal{J}(F) . \tag{2.2}
$$

The kernel in an Abelian group G on multiplication (or exponentiation) by 2 will be denoted by G_2 . Next let P_F be the set of real prime divisors of F , and $P_{A,F}$ the subset of P_F of those p for which A_p is a matrix ring or a product of matrix rings over the real quaternions $\mathbb{H}$. We define

$$\mathrm{Map}(P_{A,F}, \pm 1) = \mathrm{sign}_{A,F},$$

$$\mathrm{Map}(P_F, \pm 1) = \mathrm{sign}_F . \tag{2.3}$$

These are vector spaces over the field of 2 elements of dimension equal to the cardinality of $P_{A,F}$ (or of P_F respectively). Moreover assuming card $(P_{A,F})$ not to be zero, we have a natural embedding of the group $\{\pm 1\}$ in $\mathrm{sign}_{A,F}$, where -1 is viewed as the map which takes each $p \in P_{A,F}$ into -1 . Similarly for sign_F . If $\mathrm{sign}_{A,F} = 1$ we make the convention that $\mathrm{sign}_{A,F}/\{\pm 1\} = 1$. For the next theorem recall II Poposition 5.4(ii).

Theorem 6: _Let_ o _be global and_ $(A, \bar{\ })$ _indecomposable._

(i) _Suppose_ $(A, \bar{\ })$ _is orthogonal._ _Then_

$$\mathrm{Cok}\ \underline{P}_A \cong J(F)/J(F)^2 , \quad \mathrm{Ker}\ \underline{P}_A \cong \mathrm{sign}_{A,F}/\{\pm 1\} ,$$

and there is an exact sequence

$$1 \to (\mathrm{nrd}\ JA)_2\ /[(\mathrm{nrd}\ JA)_2 \cap \{\pm 1\}\ \mathrm{nrd}\ UA)] \to \mathrm{Ker}\ \underline{P}_A \to \mathrm{sign}_{A,F}/\{\pm 1\} \to 1 .$$

(ii) _Suppose_ $(A, \bar{\ })$ _is symplectic._ _Then_ $\underline{P}_A$ _and_ $\underline{P}_A$ _are isomorphisms._

(iii) _Suppose_ $(A, \bar{\ })$ _is unitary._ _Then_

$$\mathrm{Cok}\ \underline{P}_A \cong J(F)/N_{C/F}\ J(C) , \quad N_{C/F}\ \textit{the norm.}$$

In particular in the split case Cok $\underline{P}_A = 1$. *Also*

$\underline{P}_A$ *is injective* .

If $A = B \times B^{op}$, B *an order in a simple algebra then*

$$\mathrm{Ker}\ \underline{P}_A = \mathrm{Cl}(B) .$$

In general there is an exact sequence

$$1 \to (N_{C/F}\,\mathrm{nrd}\,A^* \cap N_{C/F}\,\mathrm{nrd}\,UA)/(N_{C/F}(\mathrm{nrd}\,A^* \cap \mathrm{nrd}\,UA)) \to \mathrm{Ker}\underline{P}_A \to \mathrm{Cl}(A)^{1-j} \to 1$$

where for notational reason we have denoted the involution $^-$ *here by* j , i.e., $\mathrm{Cl}(A)^{1-j} = [c\bar{c}^{-1} \mid c \in \mathrm{Cl}(A)]$.

Example 1) If $(A, ^-)$ is orthogonal, $A = M_n(F)$, A a maximal order, then nrd $U(A) = U(o)$ and of course $(JF)_2 \subset U(o)$. Moreover now $\mathrm{sign}_{A,F}/\{\underline{+}1\} = 1$. Thus $\underline{P}_A$, $\underline{P}_A$ are injective. The same is true more generally provided only that A_p is a matrix ring for all infinite p, and A still maximal

2) With $(A, ^-)$ orthogonal, suppose F has real prime divisors p , and for every one of these $A_p = M_n(\mathbb{H})$. If again A is the maximal order we now have

$$\mathrm{Ker}\ \underline{P}_A \cong \mathrm{sign}_{A,F}/\{\underline{+}1\} \cong \mathrm{Ker}\ \underline{P}_A .$$

Thus there are non trivial global discriminants, which are trivial everywhere locally .

3) In the case of example 1, as $\underline{P}_A$ is injective, the discriminant may now be viewed as an element of $J(F)/U(o)^2$ and the result on Cok $\underline{P}_A$ now implies that it lies in fact in $F^*J(F)^2/U(o)^2$. This then extends a theorem proved in [F1]. Moreover it implies that the image of the discriminant in $Cl(o) = J(F)/F^* U(o)$, i.e., its ideal class, is a square, generalising an old theorem of Hecke. Moreover even in the case of example 2, the same conclusion applies, again taking the image in $Cl(o)$.

4) The results on the unitary case also give new insight even in the classical case when, say, C is a quadratic extension of F, and $A = o_C$ the ring of integers in C, with $\bar{\ }$ as the automorphism of C/F. Let $Y(F)$, $Y(C)$ be the groups of global units in F, and in C. Then the exact sequence is now

$$1 \to (Y(F) \cap N_{C/F}C^*)/N_{C/F}Y(C) \to \ker \underline{P}_A \to Cl(o_C)^{1-j} \to 1 .$$

In fact this extends without change to the case $A = M_n(C)$, with A as a maximal order.

<u>Remark:</u> Cok $\underline{P}_A$ also appears as the global counterpart to the group $HCl(A)$ in the local situation, simply by going from the multiplicative group to the idele group. Recall that

$$\text{Cok } \underline{P}_A = \begin{cases} J(F)/J(F)^2, & \text{(orthogonal case)}, \\ 1 & \text{(symplectic case)}, \\ J(F)/N_{C/F}J(C) & \text{(unitary case)}, \end{cases}$$

while when o is local we have

$$
\mathrm{HCl}(A) = \begin{cases} F^*/F^{*2} & \text{(orthogonal case)}, \\ 1 & \text{(symplectic case)}, \\ F^*/N_{C/F}\,C^* & \text{(unitary case)}. \end{cases} \tag{2.4}
$$

Indeed this follows by using the "dictionary" (§1) and observing that $\mathrm{Det}\, A^* = C^*$.

<u>Proof of Theorem 6.</u> We shall first write down in the new language the expressions for the groups and maps to be considered ($\mathcal{O}$ global). We shall use the dictionary in §1, and in particular Propositions 1.1 and 1.2.

Firstly

$$
\left.\begin{aligned}
\mathrm{HCl}(A) &= F^*/(\mathrm{nrd}\, A^*)^2, & \mathrm{Ad\,HCl}(A) &= JF/(\mathrm{nrd}\, JA)^2 & &\text{(orthogonal)},\\
&= F^*/\mathrm{nrd}\, A^*, & &= JF/\mathrm{nrd}\, JA & &\text{(symplectic)},\\
&= F^*/N_{C/F}\,\mathrm{nrd}\, A^*, & &= JF/N_{C/F}\,\mathrm{nrd}\, JA & &\text{(unitary)},
\end{aligned}\right\} \tag{2.5}
$$

with $\underline{P}_A$ induced by $F^* \to JF$.

Next

$$
\left.\begin{aligned}
&\mathrm{HCl}(A) = \mathrm{Cok}\,\Delta, \quad \Delta\colon \mathrm{nrd}\, A^* \to [(\mathrm{nrd}\, JA/\mathrm{nrd}\, UA) \times F^*], \\
&\Delta(c) = (c^{-1} \bmod \mathrm{nrd}\, UA,\ c^s),\ c^s = \begin{cases} c^2 & \text{(orthogonal)}, \\ c & \text{(symplectic)}, \\ N_{C/F}c & \text{(unitary)}, \end{cases}
\end{aligned}\right\} \tag{2.6}
$$

and

$$\left.\begin{aligned} \operatorname{Ad} \operatorname{HCl}(A) &= JF/(\operatorname{nrd} UA)^2 && \text{(orthogonal)},\\ &= JF/\operatorname{nrd} UA && \text{(symplectic)},\\ &= JF/N_{C/F} \operatorname{nrd} UA && \text{(unitary)} . \end{aligned}\right\} \tag{2.7}$$

With (f mod nrd UA, g) representing an element of HCl(A), the map $\underline{P}_A$ is given by

$$\left.\begin{aligned} (f \bmod \operatorname{nrd} UA,\ g) &\mapsto f^2 g \bmod (\operatorname{nrd} UA)^2 && \text{(orthogonal)},\\ &\mapsto fg \bmod (\operatorname{nrd} UA) && \text{(symplectic)},\\ &\mapsto (N_{C/F} f) g \bmod (N_{C/F} \operatorname{nrd} UA) && \text{(unitary)} . \end{aligned}\right\} \tag{2.8}$$

(i) Orthogonal case. Clearly

$$\operatorname{Ker} \underline{P}_A = F^* \cap (\operatorname{nrd} JA)^2/(\operatorname{nrd} A^*)^2 . \tag{2.9}$$

We shall show that

$$(\operatorname{nrd} JA)^2 = (JF)^2 ,\ \text{whence}\ F^* \cap (\operatorname{nrd} JA)^2 = F^{*2} , \tag{2.10}$$

and this already implies the isomorphism for Cok $\underline{P}_A$. Certainly $(\operatorname{nrd} A_p^*)^2 = F_p^{*2}$ if p is any prime divisor outside $P_{A,F}$. If $p \in P_{A,F}$, then $\operatorname{nrd} A_p^*$ is the group of positive reals, which is its own square, i.e. $(\operatorname{nrd} A_p^*)^2 = \operatorname{nrd} A_p^* = F_p^{*2}$. Next, squaring yields an exact sequence

$$\{\pm 1\} \to F^*/\text{nrd } A^* \to F^{*2}/(\text{nrd } A^*)^2 \to 1$$

in which either $F^*/\text{nrd } A^* = 1$, or $\{\pm 1\}$ injects into it. But via the maps $c \longmapsto [p \longmapsto \text{sign}_p(c)]$, we have an isomorphism

$$F^*/\text{nrd } A^* \cong \text{sign}_{A,F} ,$$

whence finally, by (2.10),

$$F^* \cap (\text{nrd } JA)^2/(\text{nrd } A^*)^2 \cong \text{sign}_{A,F}/\{\pm 1\} . \tag{2.11}$$

By (2.9) the assertion about $\underline{P}_A$ now follows.

Next let G', G'', G''' be the following subgroups of $(\text{nrd } JA \times F^*)$.

$$G' = [(a,b), \quad a^2 b = 1] ,$$

$$G'' = [(c^{-1}h, c^2), \quad c \in \text{nrd } A^*, \quad h \in (\text{nrd } JA)_2] ,$$

$$G''' = [(c^{-1}u, c^2), \quad c \in \text{nrd } A^*, \quad u \in (\text{nrd } UA)_2] .$$

We shall establish a commutative diagram with exact rows and columns

$$
\begin{array}{ccccccccc}
 & & 1 & & 1 & & & & \\
 & & \downarrow & & \downarrow & & & & \\
1 & \to & G''' & \to & G''' & \to & 1 & & \\
 & & \downarrow & & \downarrow & & & & \\
1 & \to & G'' & \to & G' & \xrightarrow{r} & F^{*2}/(\mathrm{nrd}\ A^{*})^{2} & \to & 1 \\
 & & & & \Big\downarrow t & & & & \\
 & & & & \mathrm{Ker}\ \underline{P}_A & & & & \\
 & & & & \downarrow & & & & \\
 & & & & 1 & & & &
\end{array}
\tag{2.12}
$$

For the exact sequence of Theorem 6 (i) we then have to recall (2.10), (2.11) and to show that

$$G''/G''' \cong (\mathrm{nrd}\ JA)_2\ /[(\mathrm{nrd}\ JA)_2 \cap (\{\pm 1\}\ (\mathrm{nrd}\ UA))]\ . \tag{2.13}$$

By (2.10) the second component b for $(a,b) \in G'$ runs precisely through F^{*2} and thus r as defined by $r(a,b) = b \bmod (\mathrm{nrd}\ A^{*})^2$ is surjective. Clearly $G'' = \mathrm{Ker}\ r$, and $G''' \subset G''$. Next t is given by

$$(a,b) \mapsto (a \bmod \mathrm{nrd}\ UA,\ b)/(\mathrm{nrd}\ A^{*}) \in \mathrm{HCl}(A)\ .$$

Obviously $\mathrm{Im}\ t \subset \mathrm{Ker}\ \underline{P}_A$, and a moments thought shows that we have equality. Equally easily we see that $G''' = \mathrm{Ker}\ t$. We thus have obtained the diagram (2.12) with all required properties. Finally (2.13) (in the reverse direction) is induced from the map $(\mathrm{nrd}\ JA)_2 \to G''/G'''$ with $h \mapsto (h,1) \bmod G'''$. Clearly the map is surjective. The kernel consists of elements h with $h = c^{-1}u$, $c \in \mathrm{nrd}\ A^{*}$, $u \in (\mathrm{nrd}\ UA)$, and $c^2 = 1$, i.e. $c = \pm 1$.

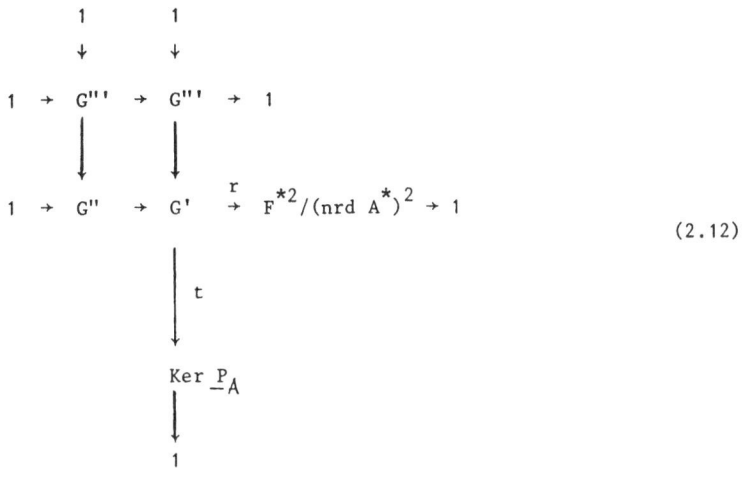

(ii) Symplectic case. That $\underline{P}_A$ is injective is the Hasse-Schilling norm theorem, that it is surjective is the result that all possible signatures are taken by elements of F^*. An element $x \in \operatorname{Ker} \underline{P}_A$ is represented by an element $(f \bmod \operatorname{nrd} UA, g)$, where $f \in \operatorname{nrd} JA$, $g \in F^*$ and $fg = 1$. Thus $g \in \operatorname{nrd} A^*$ (Hasse-Schilling) and so $x = 1$.

(iii) Unitary case. From the description given we see that $\operatorname{Im} \underline{P}_A$ consists of the classes of

$$F^*(N_{C/F} \operatorname{nrd} JA) = F^*(N_{C/F} C^*)(N_{C/F} \operatorname{nrd} JA) = F^*(N_{C/F} JF),$$

and this yields the isomorphism for $\operatorname{cok} \underline{P}_A$.

The injectivity of $\underline{P}_A$ follows from the fact that

$$F^* \cap N_{C/F} \operatorname{nrd} JA \subset N_{C/F} \operatorname{nrd} A^*. \tag{2.14}$$

To see this, note that in the unitary case $p \in P_{A,F}$ implies $A_p = M_n(\mathbb{H}) \times M_n(\mathbb{H})$, with p real and so splitting in C. One thus has to show that if $x \in N_{C/F} C^*$ and $x_p > 0$ for all $p \in P_{A,F}$, then indeed $x = N_{C/F}\, y$ with $y \in C^*$ and $y_P > 0$ for all P above all $p \in P_{A,F}$. This however is quite straightforward.

Next we show that

$$\operatorname{Ker} \underline{P}_A \cong (\operatorname{Ker} N_{C/F} \cap \operatorname{nrd} JA)/((\operatorname{Ker} N_{C/F} \cap \operatorname{nrd} A^*)(\operatorname{Ker} N_{C/F} \cap \operatorname{nrd} UA)), \tag{2.15}$$

where $N_{C/F}$ is in the first place considered as a map $J(C) \to J(F)$. A map

$$v\colon (\operatorname{Ker} N_{C/F} \cap \operatorname{nrd} JA) \to \operatorname{HCl}(A)$$

is defined as taking the element a into the class in $\mathrm{HCl}(\mathcal{A})$ of the element $(a \bmod \mathrm{nrd}\, U\mathcal{A}, 1)$. Clearly $\mathrm{Im}\, v \subset \mathrm{Ker}\, \underline{P}_{\mathcal{A}}$. On the other hand, if $(a \bmod \mathrm{nrd}\, U\mathcal{A}, b)$ represents an element in $\mathrm{Ker}\, \underline{P}_{\mathcal{A}}$ then we may suppose that $N_{C/F}\, a.b = 1$. Thus by (2.14), $b \in N_{C/F}\, \mathrm{nrd}\, A^*$ and hence indeed the given class lies in $\mathrm{Im}\, v$. Finally $a \in \mathrm{Ker}\, v$ if and only if $\exists\, c \in \mathrm{nrd}\, A^*$ with $N_{C/F}\, c = 1$, and $a = cu$, $u \in \mathrm{nrd}\, U\mathcal{A}$. But then $N_{C/F}\, u = 1$. Thus indeed we get (2.15) (in the opposite direction).

If $A = B \times B^{op}$ splits, and we have correspondingly $\mathcal{A} = \mathcal{B} \times \mathcal{B}^{op}$, then the map $x \mapsto (x, x^{-1})$ sets up isomorphisms $JF \cong \mathrm{Ker}\, N_{C/F}$, $\mathrm{nrd}\, JB \cong \mathrm{Ker}\, N_{C/F} \cap \mathrm{nrd}\, JA$, $\mathrm{nrd}\, B^* \cong \mathrm{Ker}\, N_{C/F} \cap \mathrm{nrd}\, A^*$, $\mathrm{nrd}\, U\mathcal{B} \cong \mathrm{Ker}\, N_{C/F} \cap \mathrm{nrd}\, U\mathcal{A}$, and hence an isomorphism of the right hand side in (2.15) with $\mathrm{nrd}\, JB/(\mathrm{nrd}\, B^*.\mathrm{nrd}\, U\mathcal{B}) \cong \mathrm{Cl}(\mathcal{B})$.

Finally in the general case we consider the action of the automorphism group of C/F, generated by the involution, here denoted by j. Analogously to the well known equation $H^1(JC) = 1$, one gets easily the equation $H^1(\mathrm{nrd}\, JA) = 1$. Thus $\mathrm{Ker}\, N_{C/F} \cap \mathrm{nrd}\, JA = (\mathrm{nrd}\, JA)^{1-j}$, and the surjection $\mathrm{nrd}\, JA \to \mathrm{Cl}(\mathcal{A}) = \mathrm{nrd}\, JA/(\mathrm{nrd}\, A^*)\, (\mathrm{nrd}\, U\mathcal{A})$ yields a surjection

$$\mathrm{Ker}\, N_{C/F} \cap \mathrm{nrd}\, JA \to \mathrm{Cl}(\mathcal{A})^{1-j}$$

whose kernel is $\mathrm{Ker}\, N_{C/F} \cap [(\mathrm{nrd}\, A^*)\, (\mathrm{nrd}\, U\mathcal{A})]$. By (2.15) we obtain a surjection $\mathrm{Ker}\, \underline{P}_{\mathcal{A}} \to \mathrm{Cl}(\mathcal{A})^{1-j}$ with kernel

$$\mathrm{Ker}\, N_{C/F} \cap [(\mathrm{nrd}\, A^*)(\mathrm{nrd}\, U\mathcal{A})]/[\mathrm{Ker}\, N_{C/F} \cap \mathrm{nrd}\, A^*)(\mathrm{Ker}\, N_{C/F} \cap \mathrm{nrd}\, U\mathcal{A})] .$$

To complete the proof we shall define an isomorphism from this latter group onto

$$(N_{C/F}\, \mathrm{nrd}\, A^* \cap N_{C/F}\, \mathrm{nrd}\, U\mathcal{A})/N_{C/F}\, (\mathrm{nrd}\, A^* \cap \mathrm{nrd}\, U\mathcal{A}) . \tag{2.16}$$

Let $a \in \text{nrd } A^*$, $b \in \text{nrd } UA$, $a\, b \in \text{Ker } N_{C/F}$. Map the element $a\, b \in \text{Ker } N_{C/F} \cap [(\text{nrd } A^*)\,(\text{nrd } UA)]$ onto the class of $N_{C/F}\, a = N_{C/F}\, b^{-1}$ in the group (2.16). One then verifies that this is well defined and that it induces an isomorphism.

§3. Discriminants once more

Here we shall complete the proof of Theorem 5, establishing (iii) and (iv) for indecomposable $(A, {}^-)$. We shall also add some supplementary results,referring to the case about which no explicit statement had been made in Theorem 5 namely that of $F = \mathbb{R}$ or $\mathbb{C}$.

Supplement to Theorem 5: *If* $F = \mathbb{C}$ *then* d_A *is surjective. If* $F = \mathbb{R}$ *and* $(A, {}^-)$ *is indecomposable, then* d_A *is surjective except in the following two cases (when* d_A *is null): (a)* $(A, {}^-)$ *is orthogonal,* $A = M_n(\mathbb{H})$, *(b)* $(A, {}^-)$ *is unitary,* $A = M_n(\mathbb{H}) \times M_n(\mathbb{H})$. *Here* $\mathbb{H}$ *is the quaternion division algebra.*

The proof of Theorem 5 and its supplement in the non-symplectic case will be based on

3.1 Proposition: *Let* F *be a finite extension of* $\mathbb{Q}$ *or of* $\mathbb{R}$ *or of a p-adic rational field* $\mathbb{Q}_p$. *If* $(A, {}^-)$ *is indecomposable, orthogonal or unitary, then*

$$\text{nrd } H^o(GL(A)) = H^o(\text{nrd } A^*) ,$$

where $GL(A) = \bigcup_n GL_n(A)$, *and* $H^o(X)$ *is the set of fixed points in a set* X *on which* ${}^-$ *acts by permutations.*

For the background to this in terms of the associated algebraic groups see [K] .

Remark 1: This Proposition implies the Hasse principle for nrd $H^o(GL(A))$.

Remark 2: As will be seen in the proof, our result can be refined to take account of rank. Clearly nrd $H^o(GL_n(A)) \subset$ nrd $H^o(GL_{n+1}(A))$, and in fact always nrd $H^o(GL_2(A))$ = nrd $H^o(GL(A))$. On the other hand - as we shall see - in the orthogonal case, with A a non commutative division algebra, we have nrd $H^o(GL_1(A)) \neq$ nrd $H^o(GL_2(A))$, and a similar situation obtains in some unitary cases. One can then however always derive a Hasse principle for the set nrd $H^o(GL_1(A))$. The details which are implicit in our proofs, are left to the reader. - Analogous remarks apply then also to d_A - when restricted to fixed rank.

The proofs of the last Proposition and of Theorem 5, as well as of results in the subsequent section (Theorem 7) are based on a series of Lemmas which we shall now state. These are concerned with Hermitian Morita equivalence, as defined in [FMc1] (§8) (see also [F5] §7, [Mc]), and they are in some form either stated or implicitly contained in these quoted papers. We shall briefly indicate proofs or give references. First the definition: We consider simple algebras A and B , finite dimensional over their common centre C , and involutions k of A and ℓ of B . We shall say that (A,k) and (B,ℓ) are equivalent (more precisely 1-Hermitian Morita equivalent - the more general λ-Hermitian Morita equivalence of [FMc1] will not be needed here), if there is (i) an invertible A-B-bimodule $V = {}_AV_B$ (A acting on the left, B on the right) with the actions of C via A and via B being the same (cf. [Ba1] Chapt. II) and (ii) a (always non-singular) Hermitian form $\beta: V \times V \to B$ (with respect to ℓ) so that k is the adjoint involution of β on A , i.e., so that $\beta(av_1,v_2) = \beta(v_1,a^kv_2)$, for all $a \in A$, all $v_1,v_2 \in V$. This is indeed an equivalence relation. It implies that ℓ and k have the same restriction to C . It also implies that ℓ and k are both orthogonal, or both symplectic, or both unitary - this follows as in [F5] (§7) .

3.2 Lemma: *With A,B,C as above, suppose (A,k) and (B,ℓ) are equivalent. Let h be an involution of A whose restriction to C is the same as that of k. Suppose k and ℓ are orthogonal (resp. symplectic, resp. unitary). Then each of the following statements implies the others:*

(i) *(A,h) and (B,ℓ) are equivalent.*

(ii) *There exists* $a \in A^*$ *with* $a^k = a$, *so that for all* $x \in A$, $x^h = a^{-1}x^k a$.

(iii) *h is orthogonal (resp. symplectic, resp. unitary).*

Moreover with a as under (ii) any map $h: A \to A$ *defined as in (ii) is always an involution.*

Proof: The last assertion is obvious, and the implication (i) => (iii) has been noted already.

For the implication (iii) => (ii) observe that h and k differ by an inner automorphism of A, i.e., $x^h = a^{-1}x^k a$ for some $a \in A^*$. As $h^2 = 1$, we must have $a^k = \lambda a$, $\lambda \in C^*$, $\lambda\lambda^k = 1$. If k is non-trivial on C we get $\lambda = \mu^{-1}\mu^k$, $\mu \in C^*$, and so replacing a by $a_1 = a\mu^{-1}$ we ensure that $a_1^k = a_1$. If k is trivial on C then we get $a^k = \pm a$. If $a^k = -a$ then h is symplectic (orthogonal) whenever k is orthogonal (symplectic) (see [F5] §7). Thus $a^k = a$.

Finally if (ii) holds then (A,k) and (A,h) are equivalent via the form on ${}_A A_A$ over (A,k) given by $(x,y) \mapsto x^k a y$. Thus (A,h) and (B,ℓ) are equivalent.

3.3 Lemma: (cf. [Wa2]). *Let* $A = M_n(D)$, *D a division algebra, finite dimensional over its centre C and let h be an involution of A. Then there is an involution j of D so that (A,h) and (D,j) are equivalent, except in the case when D = C and h is symplectic. In this case n = 2m and (A,h) and* $(M_2(C), \dagger)$ *are equivalent where* $\dagger$ *is the (unique) symplectic involution of* $M_2(C)$

(cf. II. (4.8)).

Proof: If $V = {}_D V_A$ is the simple A-module one shows in the standard manner that there exists a non-singular Hermitian or skew-Hermitian form γ on V over (A,h). Let ℓ be the adjoint involution of γ on D and k its matrix extension to A. Then (D,ℓ) and (A,k) are equivalent and so Lemma 3.2 becomes applicable (using implication (iii) => (i)). The restrictions of h,ℓ and k to C coincide. In the unitary case we conclude thus that (D,ℓ) and (A,h) are equivalent. Now we turn to the non-unitary case. If $D = C$ then ℓ is the identity, i.e., orthogonal. Thus if h is orthogonal then (D,ℓ) and (A,h) are equivalent. Next if $D \neq C$ then there exists an element $d \in D^*$ with $d^\ell = -d$. One of the involutions ℓ and $x \longmapsto d^{-1}x^\ell d$ of D is orthogonal, the other symplectic. Let j be that of the same type as h - i.e., orthogonal (symplectic) if h orthogonal (symplectic). The same argument as before shows that (D,j) and (A,h) are equivalent.

Finally suppose $D = C$ and h is symplectic. Then $n = 2m$ and $A = M_m(M_2(C))$, and the result follows again easily from Lemma 3.2.

3.4 Lemma: With (A,h) as in Lemma 3.3, if $a \in A^*$, $a = a^h$ then

$$a = t^h d\, t\,,$$

with $t \in A^*$ and $d = \mathrm{diag}\,(d_i)$, $d_i \in D$, a diagonal matrix.

Proof: If h is symplectic, and $D = C$, the result (with $d = 1$) is already in II. Proposition 3.1. In all other cases h is the adjoint involution of a form $\beta: D^n \times D^n \to D$, for some involution j of D. The map $x,y \longmapsto \beta(ax,y)$ $(x,y \in D^n)$ is again such a form. This can be diagonalised, i.e. there exists a basis $\{u_r\}$ of D^n so that $\beta(a\,u_r,u_s) = \delta_{rs}\, d_r$, δ_{rs} the Kronecker symbol. This yields the result.

In the next Lemma we consider algebras $B = M_r(D)$, $A = M_n(B)$, D a division algebra finite dimensional over its centre C, and involutions j on B, k on A. The matrix extension of j to A will again be denoted by j. In applying Lemma 3.4 note that a diagonal matrix of order nr with entries in D may also be viewed as a diagonal matrix of order n with entries in B. We shall write

$$\Delta: B^* = GL_1(B) \to GL_n(B) = A^*$$

for the standard embedding with

$$\Delta(b) = \operatorname{diag}(b,1,\ldots,1) .$$

3.5 Lemma: Suppose that k is given by

$$x^k = a^{-1} x^j a ,$$

with $a = a^j \in A^*$, i.e., following Lemma 3.4,

$$a = t^j d t ,$$

$t \in A^*$, $d = \operatorname{diag}(d_i)$, $d_i \in B^*$. Write

$$\Delta_t(b) = t^{-1} \Delta(b) t \quad \text{for} \quad b \in B^* .$$

Then Δ_t is an embedding $B^* \to A^*$ and

$$\Delta_t(b)^k = \Delta_t(d_1^{-1} b^j d_1) . \tag{3.1}$$

Proof: Combining the formulae for x^k and for a we get

$$t\, x^k\, t^{-1} = d^{-1}\, (t\, x\, t^{-1})^j\, d\ , \tag{3.2}$$

and hence

$$t\, \Delta_t(b)^k\, t^{-1} = d^{-1}\, \Delta(b)^j\, d.$$

But $d^{-1}\, \Delta(b)^j\, d = d^{-1}\, \Delta(b^j)\, d = \Delta(d_1^{-1}\, b^j\, d_1)$, and this gives (3.1).

Proof of Proposition 3.1. As $\mathrm{nrd}\ GL(A) = \mathrm{nrd}\ A^*$, it follows that $\mathrm{nrd}\ H^o(GL(A)) \subset H^o(\mathrm{nrd}\ A^*)$ and we are really concerned with proving the opposite inclusion. As $\mathrm{nrd}\begin{pmatrix} a_1 & 0 \\ 0 & a_2 \end{pmatrix} = \mathrm{nrd}\ (a_1)\ \mathrm{nrd}\ (a_2)$, it will suffice to show that the set $\mathrm{nrd}\ H^o(A^*)$ generates the group $H^o(\mathrm{nrd}\ A^*)$.

If $A = A_1 \times A_1^{op}$ splits then $A_1^* \cong H^o(A^*)$ as sets via $a_1 \longmapsto (a_1, a_1^{op})$, and the required result is trivial. So we shall from now on assume that $A = M_n\ (D)$, D a division algebra. As the symplectic case is excluded, it follows from Lemma 3.3 that the hypotheses of Lemma 3.5 are satisfied, with $B = D$. Using the notation of that Lemma for the involutions of A and D , we shall show that every element in $\mathrm{nrd}\ H^o(\langle j \rangle, D^*)$ is a product of elements in $\mathrm{nrd}\ H^o(\langle k \rangle, A^*)$. As $H^o(\mathrm{nrd}\ D^*) = H^o(\mathrm{nrd}\ A^*)$ this reduces the proof to the case $A = D$.

Let then $y = y^j \in D^*$. Note that the element d_1 in formula (3.1) is by hypothesis (cf. Lemma 3.4) j-symmetric, i.e., $d_1^j = d_1$. Put $y = d_1\, b$. Then $b^j\, d_1 = (d_1\, b)^j = y^j = y$, and hence $b = d_1^{-1}\, b^j\, d_1$. Therefore by Lemma 3.5., $\Delta_t(b)^k = \Delta_t(b)$. Trivially also $\Delta_t(d_1)^k = \Delta_t(d_1)$. On the other hand $\mathrm{nrd}(y) = \mathrm{nrd}(d_1)\mathrm{nrd}(b) = \mathrm{nrd}(\Delta_t(d_1))\ \mathrm{nrd}(\Delta_t(b))$, and both factors in the last product lie in $\mathrm{nrd}\ H^o(\langle k \rangle, A^*)$.

Thus from now on we take $A = D$. If $D = C$ then nrd is the identity map and again the result is trivial. So henceforth we assume that D is not a field. In the unitary case this cannot happen when $F \supset \mathbb{Q}_p$ or $F \supset \mathbb{R}$. For F a number field the result was explicitly proved by C.T.C. Wall (cf. [Wa1]) and later by C.J. Bushnell (cf. [Bu]. (When I put the problem to Bushnell and he solved it, neither I nor he realised that Wall had already settled the question). For the essential local-global step see already [K] (p. 103 Proposition a).

We are now left with the orthogonal case, and then D is a quaternion algebra. It will in fact be useful to allow here D to split, i.e., D is either a quaternion division algebra or $D = M_2(F)$. In this situation the reduced norm is given by a quaternary quadratic form. As $H^o(D)$ is a 3-dimensional subspace of D, the restriction of nrd to $H^o(D^*)$ is given by a ternary quadratic form $t = t_D$, i.e., $\mathrm{nrd}\, H^o(D^*)$ is the set T of non zero values of t. On the other hand $H^o(\mathrm{nrd}\, D^*) = \mathrm{nrd}\, D^*$. Whenever D splits then t is isotropic, i.e., universal. Hence $T = F^*$ and of course $\mathrm{nrd}\, D^* = F^*$. Next if $F = \mathbb{R}$, $D = \mathbb{H}$ then t is positive definite and T as well as $\mathrm{nrd}\, \mathbb{H}^*$ are the group of positive reals. If $F \supset \mathbb{Q}_p$ and D is a division algebra then t is an anisotropic ternary form and this implies (cf. [F4] Prop. 5) that T is the union of all but one of the cosets of F^* mod F^{*2}. This implies that every element of F^* can be written as a product of two elements of T - see below for a proof of the corresponding fact in the global case. Thus here again $\mathrm{nrd}\, H^o(D^*)$ generates $\mathrm{nrd}\, D^* = F^*$.

Finally let F be a number field. We then have a Hasse principle both for T and for $\mathrm{nrd}\, D^*$. Note that in the obvious notations $T_p = \mathrm{nrd}\, H^o(D_p^*)$. Let then $x \in \mathrm{nrd}\,(D^*)$. We shall show that $x = yz$, with $y, z \in T$. Choose $y \in F^*$ so that for each p for which D_p is a division algebra we have $y_p \in T_p$, with the further proviso that for the finite p among these we have $y_p \in F_p^{*2}$ or $y_p \notin F_p^{*2}$, according to whether $x_p \in T_p$ or $x \notin T_p$. This is possible, as for finite p we have $[F_p^* : F_p^{*2}] > 2$. Put $z = xy^{-1}$. For all p for which D splits, $z_p, y_p \in T_p = F_p^*$. For all infinite p with $D_p = \mathbb{H}$, $x_p > 0$, $y_p > 0$ hence $z_p > 0$, i.e.

$z_p, y_p \in T_p$. Next let p be a finite prime with D_p a division algebra. If $x_p \in T_p$ then $y_p \in F_p^{*2}$ and so $y_p, z_p \in T_p$. If $x_p \notin T_p$ then $z_p = x_p y_p^{-1} \not\equiv x_p \pmod{F_p^{*2}}$. As T_p only excludes one coset mod F_p^{*2}, again $z_p \in T_p$, and of course $y_p \in T_p$. Thus $y, z \in T$ as we had to show. This completes the proof of Proposition 3.1.

From Proposition 3.1, II Proposition 4.6 and our dictionary in §1 (see also (2.5)) we now get

Corollary to Proposition 3.1. *If* $(A, \bar{\ })$ *is indecomposable orthogonal, then with the identification*

$$\mathrm{HCl}(A) = F^*/(\mathrm{nrd}\ A^*)^2 ,$$

we have

$$\mathrm{Im}\ d_A = \mathrm{nrd}\ A^*/(\mathrm{nrd}\ A^*)^2 .$$

If $(A, \bar{\ })$ *is indecomposable unitary, then with the identification*

$$\mathrm{HCl}(A) = F^*/N_{C/F}\ \mathrm{nrd}\ A^* ,$$

we have

$$\mathrm{Im}\ d_A = F^* \cap \mathrm{nrd}\ A^*/N_{C/F}\ \mathrm{nrd}\ A^* .$$

Theorem 5 and its supplement - in both the orthogonal and the unitary case - are a consequence of this Corollary and of the Hasse principle for $\mathrm{nrd}\ A^*$. To complete the proof for the symplectic case we establish one more proposition.

3.6 Proposition: If $(A,\bar{\ })$ is indecomposable symplectic, then d_A is surjective.

Proof: By Lemma 3.3, $A = M_m(B)$ where B is a quaternion algebra, either a division algebra or split (i.e. $B = M_2(F)$). B has a unique symplectic involution (see II (4.8)), now to be denoted by j, and $(A,\bar{\ })$ is equivalent to (B,j). Thus, denoting by j also the matrix extension to A, we obtain by Lemma 3.2 an element $a \in A^*$, $a = a^j$ so that for all $x \in A$

$$\bar{x} = a^{-1} x^j a , \tag{3.3}$$

where by Lemma 3.4.

$$a = t^j d t, \quad d = \operatorname{diag}(d_i), \quad d_i = d_i^j \in B^* . \tag{3.4}$$

Moreover the j-symmetric elements of B are those in cent(B) = F, i.e. formula (3.1) in Lemma 3.5 now reduces to

$$\overline{\Delta_t(b)} = \Delta_t(b^j) . \tag{3.5}$$

The unique (to within equivalence) irreducible representation of B may be viewed as the embedding $B \to B \otimes_F F_c = M_2(F_c)$, and yielding an embedding $A = M_m(B) \to M_m(M_2(F_c)) = M_{2m}(F_c)$, which is in fact the irreducible representation of A. The involutions j and $\bar{\ }$ extend under these embeddings and equations (3.3) - (3.5) remain valid in this wider sense.

Using the dictionary of §1, the proof of Proposition 3.6. reduces to showing that the map $Pf^-: H^0(^-, A^*) \to F^*$ is surjective. In other words, given $c \in F^*$ we have to produce an element $y \in A^*$, $\bar{y} = y$ with $Pf^-(y) = c$. Let $b = c\, 1_B \in B^*$. Then $b^j = b$, hence by (3.5) also $\overline{\Delta_t(b)} = \Delta_t(b)$. But in $M_2(F_c)$ we see that

$$b = f^j f ,$$

where $f = \begin{pmatrix} 0 & 1 \\ -c & 0 \end{pmatrix}$. By (3.5)

$$\Delta_t(b) = \Delta_t(f^j)\,\Delta_t(f) = \overline{\Delta_t(f)}\,\Delta_t(f) ,$$

and so $Pf^-\,\Delta_t(b) = \mathrm{Det}(\Delta_t(f)) = \mathrm{Det}(f) = c$,

as we had to show. This then completes the proof of Proposition 3.6 and so of Theorem 5 and its supplement.

§4. Norm of automorphisms

We are now returning to the problem first raised in a remark after Theorem 4, in II § 6. We shall restate it here in the context of an indecomposable $(A,^-)$. As before $C = \mathrm{cent}(A)$, $F = H^0(C)$. We define

$$\rho_s': \mathrm{nrd}\, A^* \to F^*, \quad \begin{cases} x \longmapsto x^2 & \text{(orthogonal case)}, \\ x \longmapsto x & \text{(symplectic case)}, \\ x \longmapsto N_{C/F}\, x & \text{(unitary case)} . \end{cases} \tag{4.1}$$

Then

$$\mathrm{nrd}\,(\mathrm{Aut}(A,^-)) \subset \mathrm{Ker}\, \rho_s' ,$$

and the question is whether we have equality.

Theorem 7. Let F be a finite extension of $\mathbb{Q}$, or of $\mathbb{R}$, or of $\mathbb{Q}_p$ for some p. Then $\operatorname{nrd}(\operatorname{Aut}(A,\bar{\ }))=\operatorname{Ker}\rho_s'$ except in the following case, when $\operatorname{nrd}(\operatorname{Aut}(A,\bar{\ }))=1$ while $\operatorname{Ker}\rho_s'=\{\pm 1\}$: $(A,\bar{\ })$ is orthogonal, $A=M_n(D)$ with D a quaternion division algebra and either (a) $F\supset\mathbb{Q}_p$, or (b) F is a number field and $D_{\mathfrak{p}}$ splits for all infinite $\mathfrak{p}$. (For the result in the unitary case see [Wa2] (p.130) and also for the Hasse principle behind it [K] (p.103, Proposition a). For the orthogonal case see references to (4.2) below).

Proof of Theorem 7. First let $(A,\bar{\ })$ be orthogonal, $A=M_n(D)$, D a quaternion division algebra. We want to show that (cf. [Wa2] (p.129), [D] (Chapt. 2))

$$\operatorname{nrd}(\operatorname{Aut}(A,\bar{\ }))=1. \tag{4.2}$$

Let then $a\in\operatorname{Aut}(A,\bar{\ })$. Write $a=us$, with u unipotent and s semisimple in $\operatorname{Aut}(A,\bar{\ })$ (cf. [Bo] p.145). We know that $\operatorname{nrd} u=1$. Thus we may suppose that $a=s$ is semisimple. Its eigenvalues other than ± 1 occur in pairs λ,λ^{-1}. Thus it will suffice to show that the number m of eigenvalues -1 of a is even. View A as identified with $\operatorname{End}_D(D^n)$, acting on D^n from the left. The kernel space of $a+1$ is $\cong D^r$ for some r, $0\leq r\leq n$. Now extend the base field to F_c. Then $A\otimes_F F_c$ acts on $D^n\otimes_F F_c$, and the kernel space of $a\otimes 1+1\otimes 1$ is thus $\cong D^r\otimes_F F_c$, i.e. of dimension $4r$ over F_c. But $D^n\otimes_F F_c$ is the sum of two copies of the simple $A\otimes_F F_c$-module F_c^{2n}, hence the dimension of the kernel space of $a\otimes 1+1\otimes 1$ in F_c^{2n} is $2r$. As a is semisimple $m=2r$, i.e. m is even, as we had to show. We have thus established (4.2).

Still continuing with the orthogonal case, the kernel of $x\mapsto x^2$ on F^* is $\{\pm 1\}$. If D is a division algebra and either $F=\mathbb{R}$, or F is a number field and $D_{\mathfrak{p}}$ remains a division algebra for some infinite $\mathfrak{p}$, then $\{\pm 1\}\cap\operatorname{nrd} A^*=\{1\}$ and we get $\operatorname{Ker}\rho_s'=1$ and thus equality by the "backdoor". In all other cases with D a quaternion division algebra we have $\operatorname{Ker} \rho_s' = \{\pm 1\}$ and thus no equality. (This part of the proof was suggested by C.J. Bushnell).

If $A = M_n(F)$ and $(A, \bar{\ })$ is orthogonal then we can apply Lemma 3.5 with $B = F$. We see from (3.1) that $\Delta_t(-1)$ lies in $\operatorname{Aut}(A, \bar{\ })$ and its reduced norm is -1. Thus $\operatorname{Ker} \rho_s' = \operatorname{nrd} \operatorname{Aut}(A, \bar{\ })$.

Next we consider the unitary case. If first $A = A_1 \times A_1^{op}$ then $\operatorname{Ker} \rho_s'$ consists of the elements (c, c^{-1}), $c \in \operatorname{nrd} A_1^*$. If $c = \operatorname{nrd} a$, $a \in A_1^*$ then $(c, c^{-1}) = \operatorname{nrd}(a, (a^{op})^{-1}) \in \operatorname{nrd}(\operatorname{Aut}(A, \bar{\ }))$. Thus we get equality. On the other hand if $(A, \bar{\ })$ is unitary, $A = M_n(C)$, C a field quadratic over F then one can again use Lemma 3.5, with $B = C$, to establish equality, the details being more or less the same as in the orthogonal case for $A = M_n(F)$. The two types $A = A_1 \times A_1^{op}$ and $A = M_n(C)$ just discussed are the only ones which can occur in the unitary case when F is a finite extension of $\mathbb{R}$ or of $\mathbb{Q}_p$.

It remains to consider a unitary indecomposable algebra $(A, \bar{\ })$ with $A = M_n(D)$, D a division algebra, not a field, and F a number field. Here C is a quadratic extension of F with Galois group Γ. Extending the basefield we get an algebra $A \otimes_F C \cong A \times A^{op}$ with $\bar{\ }$ interchanging the two factors. By what we have proved already we have equality in this case, i.e. we get an exact sequence

$$1 \to SU(A \otimes_F C) \to U(A \otimes_F C) \xrightarrow{\operatorname{nrd}} U_+(C \otimes_F C) \to 1$$

of Γ-groups, Γ acting via the second tensor factor, where in the present context U is the unitary group, SU the special unitary group and $U_+(C \otimes_F C) = U(C \otimes_F C) \cap \operatorname{nrd}((A \otimes_F C)^*)$. (We are thus using the symbol U for the moment in a different connotation from that elsewhere). Define for each prime divisor p of F, $U_+(C_p) = U(C_p) \cap \operatorname{nrd}(A_p^*)$, and $U_+(C) = U(C) \cap \operatorname{nrd}(A^*)$. We then derive a commutative diagram with exact rows (of sets)

$$\begin{array}{ccccccccc}
1 \longrightarrow & SU(A) \longrightarrow & U(A) & \xrightarrow{\text{nrd}} & U_+(C) & \xrightarrow{\delta} & H^1(\Gamma, SU(A \otimes_F C)) \\
& \downarrow & \downarrow & & \downarrow & & \downarrow \ell \\
1 \rightarrow & \prod_p SU(A_p) \rightarrow & \prod_p U(A_p) & \xrightarrow{\text{nrd}} & \prod_p U_+(C_p) & \xrightarrow{\Pi\delta_p} & \prod_p H^1(\Gamma_p, SU(A \otimes_F C)_P) ,
\end{array}$$

where the products extend over all prime divisors p of F, and, in the last product on the right, we chose one prime divisor P of C above each p, with Γ_p the local Galois group. But we have already seen that in the local situation nrd is surjective, hence $\Pi\delta_p$ is null. But ℓ is injective (cf. [K] Chapt. V Th.1) and so δ is null, which gives the required result.

§5. Unimodular classes once more

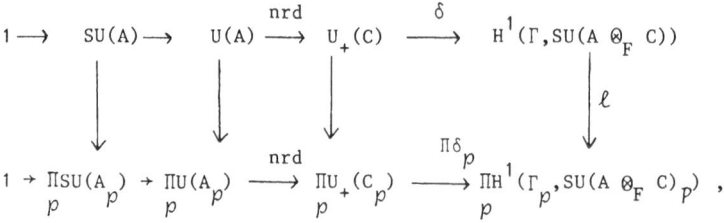

In this section we shall return to the subject matter of II §6 (specifically Propositions 6.1 - 6.2) and of II §8. Our first aim is to determine in the present context the groups Ker ν_A and Ker $\tilde{\nu}_A$. Here $\nu_A = \nu$ was defined in II. Proposition 6.1 for o local, and in II (6.14a) for o global, and $\tilde{\nu}_A = \tilde{\nu}$ was defined in II (6.14b). For o local, Ker $\nu_A = UHCl(A)$, the latter being defined in II §8, while for o global, $UHCl(A)$ is the inverse image of Ker ν_A under the map $\underline{P}_A$. This latter map has been analysed in III §2 for indecomposable involution algebras, and it is thus easy to describe $UHCl(A)$ via Ker ν_A. Our second aim is to determine explicitly the image of $UHCl(A) \rightarrow Cl(A)$, following II. Proposition 8.2. Throughout we shall pay special attention to maximal orders. For these everything reduces to indecomposable involution algebras. We shall begin by determining Det A^* (o local), Det UA (o global), for A maximal. It will be seen, as a consequence, that groups and operators which in the Hermitian theory are associated with an involution maximal order A (i.e. one with $A = \bar{A}$) have a perfectly natural definition for any maximal order and are in fact independent of the particular choice of maximal order. At the end of this section we shall then also come back briefly to the map ρ_s for maximal orders.

5.1 Proposition: Let $\mathcal{A}$ be a maximal order in the algebra A and let $\mathcal{C}$ be the maximal order of C = cent A . If $\mathcal{O}$ is local then

$$\mathrm{nrd}\, \mathcal{A}^* = \mathcal{C}^* .$$

If $\mathcal{O}$ is global then

$$\mathrm{nrd}\, U\mathcal{A} = U\mathcal{C} \cap \mathrm{nrd}\, JA .$$

We emphasize that here A need not be indecomposable, nor need it even have an involution.

Implicitly this result has already occurred earlier in these notes and is well known. To prove it we may suppose A to be simple, with centre F . Suppose first that $\mathcal{O}$ is local. Let L be a maximal non-ramified field extension of F embedded in A . Thus if $\mathcal{O}_L$ is its ring of integers then $N_{L/K}\, \mathcal{O}_L^* = \mathcal{O}^*$, $N_{L/K}$ the field norm. But L is actually a maximal subfield of A , whence $N_{L/K}\, \mathcal{O}_L^* = \mathrm{nrd}\, \mathcal{O}_L^*$. Without loss of generality, $\mathcal{A} \supset \mathcal{O}_L$, whence putting everything together $\mathrm{nrd}\, (\mathcal{A}^*) \supset \mathcal{O}^*$, and so $\mathrm{nrd}\, (\mathcal{A}^*) = \mathcal{O}^*$. The result for $\mathcal{O}$ global is obtained by putting the local results together. At infinite prime divisors $\mathfrak{p}$ the local form of the stated global equations holds trivially.

Remark 1: When A is simple, the group nrd JA is characterised inside JC by certain signature conditions at some real prime divisors $\mathfrak{p}$, namely at those at which the division algebra associated with $A_{\mathfrak{p}}$ is the real quaternion algebra $\mathbb{H}$.

Remark 2: Let for the moment $\mathcal{A}$ be a maximal order inside an involution algebra $(A,{}^{-})$ not necessarily indecomposable. For $\mathfrak{o}$ local we then have, by the Proposition, that

$$\operatorname{Det} A^{*} = \operatorname{Hom}_{\Omega_F}(K_{A,F}, (Y(F_c)) , \qquad (5.1)$$

where $Y(F_c)$ is the group of units of integers in F_c. For $\mathfrak{o}$ global we get

$$\operatorname{Det} UA = \operatorname{Det} JA \cap \operatorname{Hom}_{\Omega_F}(K_{A,F}, U(F_c)) , \qquad (5.2)$$

where $U(F_c)$ is the group of unit ideles. Again the group Det JA occurring in (5.2) represents certain signature conditions of real prime divisors. These can easily be stated explicitly, but we shall do this only for group rings, when these conditions are of a particularly simple form. We shall then obtain a nice alternative expression for the right hand side of (5.2). (See V §6).

Observe that - irrespective of whether $A = \bar{A}$ - the right hand sides in (5.1), (5.2) admit an automorphism $^{-}$ induced by the involution, and that, in the respective cases, groups $\operatorname{Hom}_{\Omega_F}(K^{s}_{A,F}, Y(F_c))$ and $\operatorname{Hom}_{\Omega_F}(K^{s}_{A,F}, U(F_c))$ are defined, and so - to give just one example - are maps

$$\operatorname{Hom}_{\Omega_F}(K_{A,F}, U(F_c)) \underset{\mathrm{Nr}}{\overset{\rho_s}{\rightleftarrows}} \operatorname{Hom}_{\Omega_F}(K^{s}_{A,F}, U(F_c)) .$$

One can thus define e.g. ν_A without assuming that $\bar{A} = A$.

From now on $(A,{}^{-})$ is again assumed to be indecomposable, and $\bar{A} = A$ is not necessarily maximal. As before $F = H^{o}(C)$, $C = \operatorname{cent} A$.

5.2 Proposition: Suppose $(A,{}^{-})$ is orthogonal

(i) *If* o *is local, then*

$$\mathrm{Ker}\ \nu_A \cong \mathrm{nrd}\ A^*/\mathrm{nrd}\ A^{*2}\ ,$$

and in particular for a maximal order A ,

$$\mathrm{Ker}\ \nu_A \cong o^*/o^{*2}\ .$$

(ii) *If* o *is global, then*

$$\mathrm{Ker}\ \tilde{\nu}_A \cong \mathrm{nrd}\ UA/\mathrm{nrd}\ UA^2\ ,$$

and in particular for a maximal order A ,

$$\mathrm{Ker}\ \tilde{\nu}_A \cong (U(o) \cap \mathrm{nrd}\ JA)/U(o)^2\ .$$

Also

$$\mathrm{Ker}\ \nu_A \cong (J_\infty(F).\mathrm{nrd}\ UA)/(J_\infty(F).\mathrm{nrd}\ UA)^2\ ,$$

and in particular for a maximal order A ,

$$\mathrm{Ker}\ \nu_A \cong U(o)/U(o)^2\ .$$

(iii) *If* o *is global, then* $\mathrm{Im}\ [U\mathrm{HCl}(A) \to \mathrm{Cl}(A)]$ *is the kernel of the composite map* $\mathrm{Cl}(A) \overset{2}{\to} \mathrm{Cl}(A) \to \tilde{\mathrm{C}}\mathrm{l}(A)$, *where the first map is*

Squaring and the second the quotient map onto the maximal quotient of $\mathrm{Cl}(A)$, *non-ramified at infinity. If* A *is a maximal, then* $\widetilde{\mathrm{Cl}}(A) = \mathrm{Cl}(\mathfrak{o})$.

The proof of this, and the following two propositions is based on a Lemma, which expresses the map Nr for indecomposable involution algebras via the dictionary of §1. Let H be a functor of fields, extended to one of finite products of fields as considered in I §2 (in particular Proposition 2.4). Here we are interested in $H(L) = L^*$ in the local case, $H(L) = J(L)$ in the global one. As before, $C = \mathrm{cent}(A)$, whence for $(A,\bar{\ })$ orthogonal or symplectic, we have $C = F$, otherwise $C \neq F$.

5.3 *Lemma: We have a commutative diagramm*

$$\begin{array}{ccc}
\mathrm{Hom}_{\Omega_F}(K^s_{A,F}, H(F_c)) & \longrightarrow & H(F) \\
\Big\downarrow \mathrm{Nr} & & \Big\downarrow t \\
\mathrm{Hom}_{\Omega_F}(K_{A,F}, H(F_c)) & \longrightarrow & H(C)
\end{array} \qquad (D)$$

where the horizontal maps are the evaluation maps as in Proposition 1.1 *and* 1.2 (see also the proof below), *and where* t *is the inclusion map for* $(A,\bar{\ })$ *orthogonal or unitary, and* $tx = x^2$ *for* $(A,\bar{\ })$ *symplectic*.

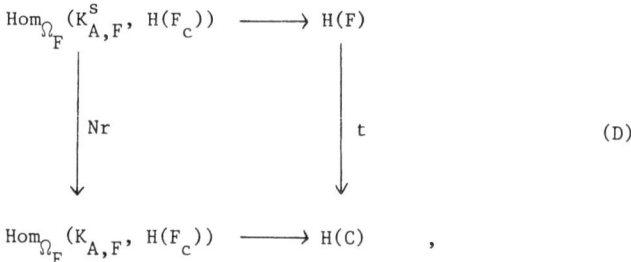

Proof: Let $\chi_1 \in K_{A,F}$ be the class of an irreducible representation. In the orthogonal and the unitary case the top row of (D) is $f \longmapsto f(\chi_1 + \bar{\chi}_1) = \mathrm{Nr}\, f(\chi_1)$. The bottom row is $g \longmapsto g(\chi_1)$ when A is a simple, and this then already shows the commutativity of (D) . When A splits the bottom row is $g \longmapsto (g(\chi_1), g(\bar{\chi}_1))$. For $g = \mathrm{Nr}\, f$ however we get $g(\chi_1) = g(\bar{\chi}_1)$ and we identify $H(F)$ with its image in $H(C)$ under the diagonal map. Thus again (D) commutes. In the symplectic case both rows are given by $f \mapsto f(\chi_1)$ and now $\mathrm{Nr}\, f(\chi_1) = f(\chi_1)^2 = tf(\chi_1)$, thus again (D) commutes.

Proof of Proposition 5.2. Let $f \in \mathrm{Det}\, A^*$. The top row of the diagram (D) in 5.3 for $H(E) = E^*$ takes f into $f(\chi_1 + \bar{\chi}_1) = f(\chi_1)^2$, the bottom row into $f(\chi_1)$. If $f = \mathrm{Det}(\mu)$, $\mu \in A^*$ then $f(\chi_1) = \mathrm{nrd}\, \mu$. Thus in terms of the right hand column of (D) the map ν_A takes the form

$$F^*/\mathrm{nrd}\, A^{*2} \to F^*/\mathrm{nrd}\, A^*$$

induced by the identity map of F^*. This in conjunction with 5.1 immediately gives (i). Analogously we derive the global formula for $\mathrm{Ker}\, \tilde{\nu}_A$. On the other hand the projection $\widetilde{K_oT(A)} \to K_oT(A)$ "coincides" with $J(F)/\mathrm{nrd}\, UA \to J(F)/J_\infty(F)\, \mathrm{nrd}\, UA$, and this yields the global formula for $\mathrm{Ker}\, \nu_A$. Note here that $J_\infty(F)^2 \subset (\mathrm{nrd}\, UA)^2$. The results for a maximal order follow by specialisation. In the local case we now have $\mathrm{nrd}\, A^* = o^*$, in the global case $\mathrm{nrd}\, UA = U(o) \cap \mathrm{nrd}\, JA$, $(J_\infty F)\,(\mathrm{nrd}\, UA) = U(o)$ and $U(o)^2 \subset (\mathrm{nrd}\, JA)^2$.

(iii) follows immediately from II. Proposition 8.2., using the dictionary of §1 and observing that in the indecomposable orthogonal (or unitary) case, $K^s_{A,F} = \mathrm{Tr}(K_{A,F})$.

Recall that we use the symbol $\hat{H}^o$ for the zero dimensional Tate cohomology group.

5.4 Proposition: Suppose that $(A, \bar{\ })$ is unitary, with C the integral closure of o in C.

(i) If o is local, then

$$\mathrm{Ker}\, \nu_A \cong \hat{H}^o(\mathrm{nrd}\, A^*) = H^o(\mathrm{nrd}\, A^*)/N_{C/F}\, \mathrm{nrd}\, A^* .$$

In particular for a maximal order A,

$$\mathrm{Ker}\ \nu_A \cong \hat{H}^0(C^*)\ .$$

(ii) If o is global, then

$$\mathrm{Ker}\ \tilde{\nu}_A \cong \hat{H}^0(\mathrm{nrd}\ UA)\ .$$

In particular for a maximal order A,

$$\mathrm{Ker}\ \tilde{\nu}_A \cong \hat{H}^0(UC \cap \mathrm{nrd}\ JA)\ .$$

Also

$$\mathrm{Ker}\ \nu_A \cong J_\infty(F)\ H^0(\mathrm{nrd}\ UA)/N_{C/F}\ \mathrm{nrd}\ UA\ .$$

In particular for a maximal order A,

$$\mathrm{Ker}\ \nu_A \cong U(o)/N_{C/F}\ (U(C) \cap \mathrm{nrd}\ JA)$$

(iii) Let o be global. Then

$$\mathrm{Im}\ [\mathcal{U}\mathrm{HCl}(A) \to \mathrm{Cl}(A)] = \mathrm{Ker}[\mathrm{Cl}(A) \xrightarrow{N_{C/F}} J(F)/(F^*.J_\infty(F).H^0(\mathrm{nrd}\ UA))]\ .$$

In particular if A is a maximal order, then

$$\mathrm{Im}\ [\mathcal{U}\mathrm{HCl}(A) \to \mathrm{Cl}(A)] = \mathrm{Ker}\ [\mathrm{Cl}(A) \to \mathrm{Cl}(\mathfrak{o})]\ .$$

Proof: The formulae for $\mathrm{Ker}\ \nu_A$ in the local case and $\mathrm{Ker}\ \tilde{\nu}_A$ in the global one are established in the same way as those in Proposition 5.2, i.e., via Lemma 5.3, and similarly for $\mathrm{Ker}\ \nu_A$, when $\mathfrak{o}$ is global. Indeed, to illustrate this for $\mathfrak{o}$ global, using the dictionary of §1, we get maps

$$J(F)/N_{C/F}\ \mathrm{nrd}\ UA \to J(C)/\mathrm{nrd}\ UA \to J(C)/J_\infty(C)\ \mathrm{nrd}\ UA\ ,$$

the first map on the left being $\tilde{\nu}_A$ and the composition of the two maps being ν_A. This implies the stated isomorphisms. The special formulae for maximal orders are deduced using the known descriptions $\mathrm{nrd}\ A^* = C^*$ ($\mathfrak{o}$ local), $\mathrm{nrd}\ UA = UC \cap \mathrm{nrd}\ JA$, and the fact that $J_\infty(C)\ (\mathrm{nrd}\ JA) = J(C)$. Finally (iii) follows now from II Proposition 8.2. on translating.

Now we turn to the symplectic case, where diagram (D) in Lemma 5.2 looks different and where $K^s_{A,F} \neq \mathrm{Tr}(K_{A,F}) = 2K_{A,F} = 2K^s_{A,F}$, and hence II Proposition 8.2. becomes less precise. In the local case $\mathrm{nrd}\ A^*$ is an open subgroup of $\mathfrak{o}^*$, hence

$$V_A = [x \in F^* \mid x^2 \in \mathrm{nrd}\ A^*] \tag{5.3}$$

is an open subgroup of $\mathfrak{o}^*$. Similarly for $\mathfrak{o}$ global $\mathrm{nrd}\ UA$ is an open subgroup of $U(\mathfrak{o})$, hence

$$V_A = [x \in J(F) \mid x^2 \in \mathrm{nrd}\ UA] \tag{5.4}$$

is an open subgroup of $U(\mathfrak{o})$.

5.5 Proposition: Suppose $(A,\bar{\ })$ is symplectic

(i) If o is local, then

$$\mathrm{Ker}\ \nu_A \cong V_A/\mathrm{nrd}\ A^* .$$

In particular if A is a maximal order,

$$\mathrm{Ker}\ \nu_A = 1 .$$

(ii) If o is global, then

$$\mathrm{Ker}\ \tilde{\nu}_A = \mathrm{Ker}\ \nu_A \cong V_A/\mathrm{nrd}\ UA .$$

In particular if A is a maximal order,

$$\mathrm{Ker}\ \nu_A = U(o)/\mathrm{nrd}\ UA .$$

(iii) If o is global, then

$$\mathrm{Im}\ [UHCl(A) \to Cl(A)] = \mathrm{Ker}[Cl(A) = JF/(F^*.\mathrm{nrd}\ UA) \to JF/F^*.V_A] .$$

In particular for a maximal order A,

$$\mathrm{Im}\ [UHCl(A) \to Cl(A)] = \mathrm{Ker}[Cl(A) \to Cl(o)] .$$

Proof: We consider the global case, the local one following by the same method. The horizontal maps in (D) (Lemma 5.3) are both isomorphisms with $JF/\mathrm{nrd}\ UA$, and thus by the Lemma, $\tilde{\nu}_A$ is $J(F)/\mathrm{nrd}\ UA \xrightarrow{2} J(F)/\mathrm{nrd}\ UA$ (squaring). ν_A is the composition of this with the quotient map $J(F)/\mathrm{nrd}\ UA \to J(F)/J_\infty(F).\mathrm{nrd}\ UA$. Thus indeed $\mathrm{Ker}\ \tilde{\nu}_A = V_A/\mathrm{nrd}\ UA$. If now the class of $x \bmod \mathrm{nrd}\ UA$ lies in $\mathrm{Ker}\ \nu_A$, then $x^2 \in J_\infty(F)\ \mathrm{nrd}\ UA$. Write $x = x_\infty x_{fin}$ as a product of an idele $x_\infty \in J_\infty(F)$ and a finite idele x_{fin}. Then $x_{fin}^2 \in \mathrm{nrd}\ UA$, $x_\infty^2 \in J_\infty(F)^2 \subset \mathrm{nrd}\ UA$, so $x^2 \in \mathrm{nrd}\ UA$, and $x \in V_A$, as we wanted to show. Thus $\mathrm{Ker}\ \nu_A = V_A/\mathrm{nrd}\ UA$. Next for (iii), let $(a, b) \in \mathrm{nrd}\ JA \times F^*$. Then its class in $\mathrm{HCl}(A)$ lies in $U\mathrm{HCl}(A)$ precisely if $(a\ b)^2 = u \in \mathrm{nrd}\ UA$, i.e., $ab \in V_A$, i.e. $a \in F^* V_A$. This gives the result.

Finally the assertions for a maximal order A follow from the fact that in this case $\mathrm{nrd}\ A^* = V_A = o^*$ (o local), and $V_A = U(o)$ (o global).

Our last result deals with the restriction map ρ_s for maximal orders. For the moment consider a, not necessarily indecomposable, $(A, \bar{\ })$ with a maximal order $A = \bar{A}$. If e.g. o is local then in view of (5.1) the map ρ_s which takes $\mathrm{Det}\ A^*$ into $\mathrm{Det}^s\ A^*$ may be viewed instead as going into $\mathrm{Hom}_{\Omega_F}(K^s_{A,F}, Y(F_c))$, i.e. we consider

$$\rho_s\colon \mathrm{Hom}_{\Omega_F}(K_{A,F}, Y(F_c)) \to \mathrm{Hom}_{\Omega_F}(K^s_{A,F}, Y(F_c)) \tag{5.5}$$

and we shall want to find $\mathrm{Cok}\ \rho_s$. Incidentally this map is defined without any reference to an involution stable maximal order A. In view of (5.2), analogous remarks apply also when o is global. In this case our map is

$$\rho_s\colon \mathrm{Hom}_{\Omega_F}(K_{A,F}, U(F_c)) \cap \mathrm{Det}\ JA \to \mathrm{Hom}_{\Omega_F}(K^s_{A,F}, U(F_c)) \cap \mathrm{Det}^s JA\,. \tag{5.6}$$

Now we shall again assume that $(A, \bar{\ })$ is indecomposable, and use our dictionary. We then get, by Propositions 1.1 and 1.2,

5.6 Proposition: Let A be a maximal order.

(i) Suppose $(A,\bar{\ })$ is orthogonal. If o is local, then

$$\mathrm{Cok}\ \rho_s \cong o^*/o^{*2} .$$

If o is global, then

$$\mathrm{Cok}\ \rho_s \cong (U(o) \cap \mathrm{nrd}\ JA)/(U(o) \cap \mathrm{nrd}\ JA)^2 = (U(o) \cap \mathrm{nrd}\ JA)/U(o)^2 .$$

(ii) Suppose $(A,\bar{\ })$ is unitary and C the integral closure of o in $C = \mathrm{cent}(A)$. If o is local, then

$$\mathrm{Cok}\ \rho_s \cong o^*/N_{C/F}\ C^* = \hat{H}^o(C^*) .$$

If o is global, then

$$\mathrm{Cok}\ \rho_s \cong \hat{H}^o(U(C) \cap \mathrm{nrd}\ JA) .$$

(iii) Suppose that $(A,\bar{\ })$ is symplectic. Then ρ_s is the identity map.

CHAPTER IV CHANGE OF ORDER

Here we shall give a unified treatment both for "going up" and for "going down". Even where no involution is involved, our approach throws new light on the problem, and it is in this context that the superiority of the Hom group language is most marked. Although many results, and in particular parts of Theorems 8 and 10 are more general, we shall make again the blanket assumptions that $\mathfrak{o}$ is global or local or a field - as in the preceding chapters. This is preferable to having at each stage lengthy detailed statements of the precise conditions under which some assertion holds. The "generality" game is left to the reader, be he so inclined.

With some significant exceptions, the change of order maps commute with the homomorphisms defined at various places in the preceding chapters. We shall not state some of the more obvious such naturality results.

§1. Going up

We consider homomorphisms $\Sigma: A \to B$ of finite dimensional, separable F-algebras and compatible homomorphisms $\Sigma: \mathfrak{A} \to \mathfrak{B}$ of orders spanning these, with associated covariant homomorphisms of Grothendieck groups and classgroups. These will be seen to correspond to contravariant homomorphisms on the groups $K_{A,F}$ etc.

Transition from $\mathfrak{A}$ to A, and local completions $\mathfrak{A} \to \mathfrak{A}_{\mathfrak{p}}$, $A \to A_{\mathfrak{p}}$ have already been dealt with (localisation without completion can be treated in the same manner, but is of no interest to us). These are the only important cases where the base ring $\mathfrak{o}$ cannot be considered as fixed. In the sequel we shall accordingly restrict ourselves to homomorphisms Σ over a fixed base ring $\mathfrak{o}$ (and so a fixed basefield F). Examples: (i) $A = B$ with $\mathcal{A} \subset \mathcal{B}$. (ii) $A = F\Delta$, $B = F\Gamma$ are group rings of finite groups Δ and Γ, with a homomorphism $\Delta \to \Gamma$ given, e.g. Δ a subgroup of Γ or Γ a quotient of Δ. (iii) E is a finite extension field of F with $\mathcal{O}_E$ the integral closure of $\mathcal{O}$, and for given $\mathcal{A}$, A, we have $B = A \otimes_F E$, $\mathcal{B} = \mathcal{A} \otimes_{\mathcal{O}} \mathcal{O}_E$.

We first give the results which do not involve an involution.

Theorem 8. (i) Σ gives rise to additive functors $W \longmapsto W \otimes_{\mathcal{A}} \mathcal{B}$ on the categories of projectives, of locally free, and of l.f.p. torsion modules and induces homomorphisms of sequences I (1.1), I (1.2) and of the diagram in Theorem 1.

(ii) Associating with a representation $T: B \to M_n(F_c)$ over F the representation $\Sigma'T = T \circ \Sigma : A \to M_n(F_c)$, we obtain a homomorphism $\Sigma': K_{B,F} \to K_{A,F}$ of Ω_F-modules, and hence a homomorphism

$$\Sigma : \mathrm{Hom}_{\Omega_F}(K_{A,F}, \cdot) \to \mathrm{Hom}_{\Omega_F}(K_{B,F}, \cdot) . \tag{1.1}$$

(iii) With L a commutative F-algebra, extend Σ to $\Sigma \otimes 1 : A \otimes_F L \to B \otimes_F L$. Then the diagram

$$\begin{array}{ccc} GL_q(A \otimes_F L) & \xrightarrow{\Sigma \otimes 1} & GL_q(B \otimes_F L) \\ \downarrow \mathrm{Det} & & \downarrow \mathrm{Det} \\ \mathrm{Hom}_{\Omega_F}(K_{A,F}, (F_c \otimes_F L)^*) & \xrightarrow{\Sigma} & \mathrm{Hom}_{\Omega_F}(K_{B,F}, (F_c \otimes_F L)^*) \end{array} \tag{1.2}$$

commutes.

(iv) The maps (1.1) take Det $J\mathcal{A}$, Det $U\mathcal{A}$, (for $\mathcal{O}$ global) and

$\text{Det}\ A^*$, $\text{Det}\ \mathcal{A}^*$ *into* $\text{Det}\ JB$, $\text{Det}\ U\mathcal{B}$, $\text{Det}\ B^*$, $\text{Det}\ \mathcal{B}^*$, *respectively. The isomorphisms of Theorem 2 commute with the maps induced by* Σ .

Proof: The functorial property for projectives or for locally free modules is trivial. Consider then an exact sequence

$$0 \to Y \to X \to M \to 0$$

of $\mathcal{A}$-modules, with X, Y locally free and M l.f.p. torsion. We get an exact sequence of $\mathcal{B}$-modules

$$0 \to \text{Tor}_{\mathcal{A}}(M,\mathcal{B}) \to Y \otimes_{\mathcal{A}} \mathcal{B} \to X \otimes_{\mathcal{A}} \mathcal{B} \to M \otimes_{\mathcal{A}} \mathcal{B} \to 0 \quad .$$

Certainly $\text{Tor}_{\mathcal{A}}(M,\mathcal{B})$ and $M \otimes_{\mathcal{A}} \mathcal{B}$ are $\mathfrak{o}$-torsion modules, $Y \otimes_{\mathcal{A}} \mathcal{B}$, $X \otimes_{\mathcal{A}} \mathcal{B}$ are locally free $\mathcal{B}$-modules. Hence, as $\text{Tor}_{\mathcal{A}}(M,\mathcal{B})$ is a submodule of the torsion free $\mathfrak{o}$-module $Y \otimes_{\mathcal{A}} \mathcal{B}$, it is null. Thus indeed $M \otimes_{\mathcal{A}} \mathcal{B}$ is l.f.p. torsion over $\mathcal{B}$, and $M \longmapsto M \otimes_{\mathcal{A}} \mathcal{B}$ is an exact functor. We now already get the appropriate change of order homomorphisms on the classgroups and K_o-groups and the fact that they commute with all the relevant maps as asserted under (i), except possibly the map θ of Theorem I (ii). For the latter we have to establish the commutativity of the diagram

$$\begin{array}{ccc} GL_n(\mathcal{A}) & \xrightarrow{\theta_n} & K_o\ T\mathcal{A} \\ \downarrow{\scriptstyle \Sigma} & & \downarrow{\scriptstyle \Sigma} \\ GL_n(\mathcal{B}) & \xrightarrow{\theta_n} & K_o\ T\mathcal{B} \end{array} \qquad (1.3)$$

,

and then Abelianize and go to the limit on the left hand column. (θ_n was defined prior to Theorem 1.) Indeed let X, Y be locally free A-modules spanning the same A-module V of rank n . Choosing an A-basis $\{v_i\}$ of V , with corresponding B-basis $\{v_i \otimes 1\}$ of $V \otimes_A B$, we get the vertical isomorphisms in the commutative diagram

$$\begin{array}{ccc} \mathrm{Aut}_A(V) & \xrightarrow{\alpha \mapsto \alpha \otimes 1} & \mathrm{Aut}_B(V \otimes_A B) \\ \wr\| & & \wr\| \\ GL_n(A) & \xrightarrow{\Sigma} & GL_n(B) \end{array} \qquad (1.4)$$

Assume, without loss of generality, that $Y = \alpha X \subset X$, and that α maps to $a \in GL_n(A)$. Then $\theta_n(a) = [X/Y]_A$. But $Y \otimes_A B = (\alpha \otimes 1)(X \otimes_A B)$ and so $\theta_n(\Sigma a) = [X \otimes_A B/Y \otimes_A B]_B$, as we had to show.

For (ii) we only have to note that $T \mapsto \Sigma' T$ preserves sums and equivalence. For (iii) let $a \in GL_q(A \otimes_F L)$, and let χ be an element of $K_{B,F}$ corresponding to an actual representation T . Then $\Sigma\, \mathrm{Det}(a)(\chi) = \mathrm{Det}(a)(\Sigma'\chi) = \mathrm{Det}(T o \Sigma)(a) = \mathrm{Det}\, T(\Sigma a) = \mathrm{Det}(\Sigma a)(\chi)$ and this shows that the diagram (1.2) does indeed commute.

The fact that the maps (1.1) have the required effect on Det JA etc. follows now from the fact that Σ induces maps $JA \to JB$ etc. and that (1.2) commutes. That isomorphisms (i) and (ii) of Theorem 2 "commute" with Σ follows again by the commutativity of (1.2) on going to the limit. For the other isomorphisms of Theorem 2 one then has to use (the already established) part (i) of the Theorem 8 as well.

Corollary: *If* A = B *then the maps*

$$K_o T(A) \to K_o T(B), \quad Cl(A) \to Cl(B), \quad K_o(A) \to K_o(B)$$

<u>are surjective</u>.

<u>Remark</u>: We consider the case $B = A \otimes_F E$, with E a finite extension field of F . From I, Corollary to Proposition 2.1 and Proposition 2.2 we get an isomorphism

$$\alpha\colon \mathrm{Hom}_{\Omega_F}(K_{B,F}, X) \cong \mathrm{Hom}_{\Omega_E}(K_{B,E}, X) , \quad \text{for } \Omega_F\text{-modules } X .$$

It can also be described as the map induced by the inclusion $K_{B,E} \subset K_{B,F}$. On the other hand by I. Proposition 2.3, we get an isomorphism

$$\beta\colon K_{B,E} \cong K_{A,F} ,$$

given by restriction to A . If χ is the class of a representation T of A , then $\beta^{-1}\chi$ is that of the representation $(a \otimes e) \longmapsto T(a)\, e$. We know that on composing representations with operations of Galois groups, $K_{B,E}$ acquires the structure of an Ω_E-module, $K_{A,F}$ that of an Ω_F-module. Now use β to define on $K_{B,E}$ the structure of an Ω_F-module. One verifies that the resulting two structures of Ω_E-module on $K_{B,E}$ coincide. Thus we get a diagram

$$\begin{array}{ccc} \mathrm{Hom}_{\Omega_F}(K_{A,F}, \cdot) & \xrightarrow{\Sigma} & \mathrm{Hom}_{\Omega_F}(K_{B,F}, \cdot) \\ \downarrow \wr \beta & & \downarrow \wr \alpha \\ \mathrm{Hom}_{\Omega_F}(K_{B,E}, \cdot) & \rightarrow & \mathrm{Hom}_{\Omega_E}(K_{B,E}, \cdot) , \end{array} \qquad (1.5)$$

where the bottom row is the (obvious) inclusion map. One then verifies easily

1.1 *Proposition: Diagram* (1.5) *commutes.*

This means that in this situation the classgroup maps are essentially induced by the embedding $\mathrm{Hom}_{\Omega_F} \subset \mathrm{Hom}_{\Omega_E}$.

From now on we shall assume that $\Sigma\colon (A,\bar{\ }) \to (B,\bar{\ })$ is a homomorphism of involution algebras and that $\bar{A} = A$, $\bar{B} = B$.

Theorem 9. (i) *The maps induced by* Σ (cf. Theorem 8) *commute with the automorphisms* $\bar{\ }$.

(ii) $\Sigma'\colon K_{B,F} \to K_{A,F}$ *induces a homomorphism* $K^s_{B,F} \to K^s_{A,F}$, *hence homomorphisms*

$$\Sigma\colon \mathrm{Hom}_{\Omega_F}(K^s_{A,F}, \cdot) \to \mathrm{Hom}_{\Omega_F}(K^s_{B,F}, \cdot) .$$

(iii) *The diagrams*

$$\begin{array}{ccc} H^o(GL_n(A)) & \xrightarrow{\Sigma} & H^o(GL_n(B)) \\ \downarrow{\scriptstyle Pf} & & \downarrow{\scriptstyle Pf} \\ \mathrm{Hom}_{\Omega_F}(K^s_{A,F}, F^*_c) & \longrightarrow & \mathrm{Hom}_{\Omega_F}(K^s_{B,F}, F^*_c) \end{array} \qquad (1.6)$$

commute.

(iv) *Associating with a Hermitian* A*-module* (X,h) *the Hermitian* B*-module* $(X \otimes_A B,\ h \otimes_A B)$, *yields a functor, preserving* *orthogonal sums, hence a homomorphism* $K_0H(\mathcal{A}) \to K_0H(\mathcal{B})$. *Moreover* Σ *induces homomorphisms*

$$HCl(\mathcal{A}) \to HCl(\mathcal{B}), \quad Hcl(A) \to HCl(B)$$

and for $\mathcal{O}$ *global*

$$\mathrm{Ad}\ HCl(\mathcal{A}) \to \mathrm{Ad}\ HCl(\mathcal{B}), \quad \mathrm{Ad}\ HCl(A) \to \mathrm{Ad}\ HCl(B) .$$

(v) *We have a commutative diagram*

$$\begin{array}{ccc} K_0\ H(\mathcal{A}) & \longrightarrow & K_0\ H(\mathcal{B}) \\ \Big\downarrow d & & \Big\downarrow d \\ HCl(\mathcal{A}) & \longrightarrow & HCl(\mathcal{B}) \end{array} \qquad (1.7)$$

and similar other diagrams involving the maps considered in II §5, 6.

(vi) Σ *induces a homomorphism* $\widetilde{K_0\ H}(\mathcal{A}) \to \widetilde{K_0\ H}(\mathcal{B})$ *and homomorphisms of sequences* II (5.16) *and* II (5.17), *the map* $\mathbb{Z} \to \mathbb{Z}$ *being the identity.*

Proof: The map $\Sigma': K_{B,F} \to K_{A,F}$ clearly preserves $^-$ action, hence so does the map Σ on the Hom groups. Using now Theorem 8, Theorem 2 and II Proposition 1.3 we deduce the same for $K_0\ T(\mathcal{A}) \to K_0\ T(\mathcal{B})$, $K_0(\mathcal{A}) \to K_0(\mathcal{B})$ etc. (ii) is obvious. - Next the proof of commutativity for (1.6) is analogous to that for (1.2) in Theorem 8. - From (ii) and Theorem 8 one easily deduces that Σ induces maps on the various Hermitian classgroups. The assertion on the functor from Hermitian $\mathcal{A}$-modules to Hermitian $\mathcal{B}$-modules and the resulting map on K_0H is obvious. For the commutativity of (1.7) one uses that of (1.2) and (1.6), following the same type of argument as that used in the proof of Theorem 8 (cf. (1.4)). The other assertions in (v) and (vi) are straightforward to verify.

Remark 1. Preservation of $^{-}$ action for K_0-groups can also be deduced from natural isomorphisms such as

$$\mathrm{Hom}_{\mathcal{A}}(X,\mathcal{A}) \otimes_{\mathcal{A}} \mathcal{B} \cong \mathrm{Hom}_{\mathcal{B}}(X \otimes_{\mathcal{A}} \mathcal{B}, \mathcal{B}),\ X \text{ locally free on } \mathcal{A}.$$

Remark 2. There is an analogue to Proposition 1.1 for K^s.

§2. Going down

Here we consider restriction of scalars for modules, which in terms of the groups $K_{A,F}$, $K^s_{A,F}$ corresponds to induction. Throughout this section A and B are finite dimensional, separable F-algebras, with $B \subset A$, and $\mathcal{A}$ and $\mathcal{B}$ are $\mathcal{O}$-orders spanning A and B respectively, with $\mathcal{B} \subset \mathcal{A}$ and $\mathcal{A}$ locally free as right $\mathcal{B}$-module. Then A is free as a B-module. Throughout we shall write m for its rank. Two examples: (i) E is a finite extension field of F and $A = B \otimes_F E$, (ii) $B = F\Delta$, $\mathcal{B} = \mathcal{O}\Delta$, $A = F\Gamma$, $\mathcal{A} = \mathcal{O}\Gamma$ with Δ a subgroup of a finite group Γ.

We shall again first deal with the problem without considering involutions.

Theorem 10. (i) Restriction of scalars from $\mathcal{A}$ to $\mathcal{B}$ gives rise to additive functors on the categories of locally free and of l.f.p. torsion modules and thus to homomorphisms $\Lambda : K_0T(\mathcal{A}) \to K_0T(\mathcal{B})$, and $K_0(\mathcal{A}) \to K_0(\mathcal{B})$, $\mathrm{Cl}(\mathcal{A}) \to \mathrm{Cl}(\mathcal{B})$ which commute with sequences

I (1.1), (where now $\mathbb{Z} \to \mathbb{Z}$ is multiplication by m). Analogously for projectives.

(ii) Left action of $\mathcal{A}$ on the right $\mathcal{B}$-module $\mathcal{A}_{\mathcal{B}}$ induces an embedding

$$\Lambda : \mathcal{A} \to \mathrm{End}(\mathcal{A}_{\mathcal{B}}) .$$

Moreover $\mathcal{A}_{\mathcal{B}}$ being locally free it is a $\mathcal{B}$-progenerator, i.e., $\mathrm{End}(\mathcal{A}_{\mathcal{B}})$ is Morita equivalent to $\mathcal{B}$ (cf. [Ba1] Chapt. II), hence, via left action of $\mathcal{A}$ on $\mathcal{A}$, we get a homomorphism

$$\Lambda: K_1(\mathcal{A}) \to K_1(\mathrm{End}(\mathcal{A}_{\mathcal{B}})) = K_1(\mathcal{B}) ,$$

and analogously

$$\Lambda: K_1(A) \to K_1(B) .$$

The maps Λ under (i), (ii) yield a homomorphism of the diagram in Theorem 1. (ii).

(iii) Choosing a basis of A_B over B, and so an isomorphism $\mathrm{End}(A_B) \cong M_m(B)$, we obtain from the left action of A on A_B an embedding

$$\Lambda: A \to M_m(B) ,$$

and hence, for every representation $T: B \to M_n(F_c)$ over F, a representation

$$\Lambda' T = T \circ \Lambda : A \to M_{mn}(F_c) \quad \text{over} \quad F .$$

There results a homomorphism $\Lambda' : K_{B,F} \to K_{A,F}$ *of* Ω_F*-modules, independent of the above choice of basis, and hence a homomorphism*

$$\Lambda : \mathrm{Hom}_{\Omega_F}(K_{A,F}, \cdot) \to \mathrm{Hom}_{\Omega_F}(K_{B,F}, \cdot) .$$

(iv) *If* L *is a commutative* F-*algebra then the diagram*

$$\begin{array}{ccc} GL_n(A \otimes_F L) & \xrightarrow{\Lambda} & GL_{nm}(B \otimes_F L) \\ \downarrow \mathrm{Det} & & \downarrow \mathrm{Det} \\ \mathrm{Hom}_{\Omega_F}(K_{A,F}, (F_c \otimes_F L)^*) & \xrightarrow{\Lambda} & \mathrm{Hom}_{\Omega_F}(K_{B,F}, (F_c \otimes_F L)^*) \end{array}$$

commutes.

(v) *The homomorphisms* Λ *on Hom groups* (cf. (iii)) *induce homomorphisms*

$$\mathrm{Det}\, A^* \to \mathrm{Det}\, B^* ,$$

$$\mathrm{Det}\, \mathcal{A}^* \to \mathrm{Det}\, \mathcal{B}^* , \qquad (\mathfrak{o} \text{ local}),$$

$$\mathrm{Det}\, JA \to \mathrm{Det}\, JB , \quad \mathrm{Det}\, U\mathcal{A} \to \mathrm{Det}\, U\mathcal{B} \qquad (\mathfrak{o} \text{ global}) .$$

The isomorphisms of Theorem 2 *then commute with the homomorphisms induced by* Λ , *and diagrams such as*

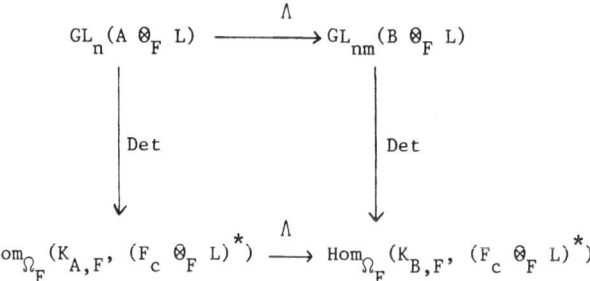

$$\begin{array}{ccccc}
\mathrm{Det}\ UA & \to & \mathrm{Det}\ JA & \to & K_oT(A) \\
\downarrow & & \downarrow & & \downarrow \\
\mathrm{Det}\ UB & \to & \mathrm{Det}\ JB & \to & K_oT(B)
\end{array}$$

will commute.

Proof: (i) only involves forgetful functors and is obvious. The maps on K_1 in (ii) are standard. The only part of (ii) which requires some explanation is the commutativity of

$$\begin{array}{ccc}
K_1(A) & \text{———} & K_oT(A) \\
| & & | \\
K_1(B) & \text{———} & K_oT(B)
\end{array}$$

which will follow from the commutativity of the diagram

$$\begin{array}{ccc}
GL_n(A) & \xrightarrow{\theta_{n,A}} & K_oT(A) \\
\downarrow \Lambda & & \downarrow \Lambda \\
GL_{nm}(B) & \xrightarrow{\theta_{nm,B}} & K_oT(B)
\end{array} \qquad (2.1)$$

with θ defined prior to Theorem 1. To establish this, consider a free A-module V of rank n. We clearly have a commutative diagram

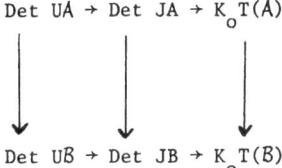

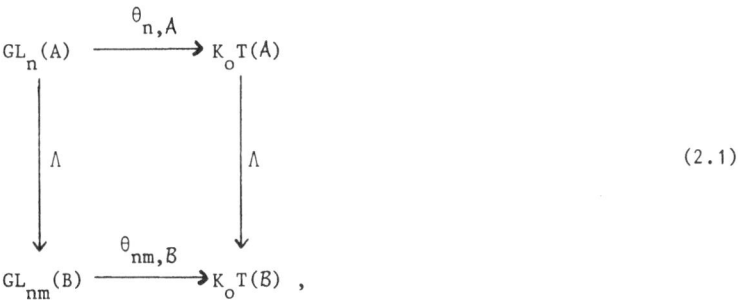

$$\begin{array}{ccc} \mathrm{Aut}(V_A) & \longrightarrow & K_oT(A) \\ \downarrow & & \downarrow \Lambda \\ \mathrm{Aut}(V_B) & \longrightarrow & K_oT(B) \end{array} ,$$

where the left hand column is just the inclusion map of the group of automorphisms of V as A-module into that of automorphisms of V as B-module. The rows are $\alpha \mapsto [X/\alpha X]_A$, $\alpha \mapsto [X/\alpha X]_B$, with X free, say under the hypothesis that $\alpha X \subset X$. But if we fix a free A-basis $\{v_i\}$ of V, then, for a given B-basis $\{a_r\}$ of A_B, we get a free B-basis $\{v_i a_r\}$ of V, and via these bases we get a commutative diagram

$$\begin{array}{ccc} GL_n(A) & \longrightarrow & \mathrm{Aut}(V_A) \\ \Lambda \downarrow & & \downarrow \\ GL_{nm}(B) & \longrightarrow & \mathrm{Aut}(V_B) \end{array}$$

and (2.1) is obtained by composing this with the preceding diagram above.

(iii) is obvious. For (iv) let $a \in GL_n(A \otimes_F L)$ and let χ be an element of $K_{B,F}$ corresponding to a representation $T\colon B \to M_r(F_c)$. Then

$$(\Lambda\ \mathrm{Det}(a))(\chi) = \mathrm{Det}(a)(\Lambda'\chi)$$

$$= \mathrm{Det}\ \Lambda'T(a) = \mathrm{Det}\ T(\Lambda(a)) = (\mathrm{Det}\ \Lambda(a))(\chi) ,$$

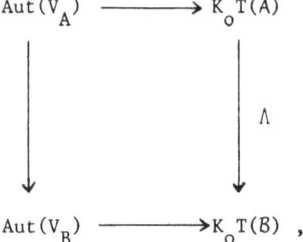

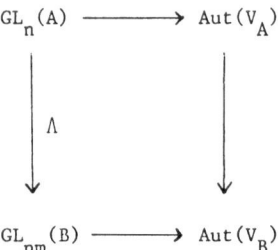

as we had to show.

Under (v) the fact that Λ maps Det A^* into Det B^*, and that it commutes with the isomorphism (i) of Theorem 2 is a consequence of (iv). Analogously one gets, for $\mathcal{O}$ local, the fact that Λ maps Det $\mathcal{A}^*$ into Det $\mathcal{B}^*$, and that it commutes with the isomorphism (ii) of Theorem 2. From (i) and (ii) one now deduces that Λ commutes with the isomorphism (iii) in Theorem 2, i.e., that the diagram

$$\begin{array}{ccccccccc} 0 & \to & \mathrm{Det}\,\mathcal{A}^* & \to & \mathrm{Det}\,A^* & \to & K_oT(\mathcal{A}) & \to & 0 \\ & & \downarrow & & \downarrow & & \downarrow & & \\ 0 & \to & \mathrm{Det}\,\mathcal{B}^* & \to & \mathrm{Det}\,B^* & \to & K_oT(\mathcal{B}) & \to & 0 \end{array}$$

commutes (for $\mathcal{O}$ local).

The global results in turn follow by localisation and by what has already been shown.

We shall next consider the following special case:

$$A = B \otimes_F E, \quad E \text{ a finite extension field of } F, \ E \subset F_c\,. \tag{2.2}$$

Thus Ω_E is of finite index in Ω_F. We therefore have the usual trace map on homomorphisms. As we are using multiplicative notation we shall actually call it a <u>norm map</u>. It is denoted by

$$\mathcal{N}_{E/F}: \mathrm{Hom}_{\Omega_E}(G,H) \to \mathrm{Hom}_{\Omega_F}(G,H) \tag{2.3}$$

with Ω_F-modules G and H, and defined by

$$(\mathcal{N}_{E/F}f)(g) = \prod_\sigma f(g^{\sigma^{-1}})^\sigma ,$$

where $\{\sigma\}$ is a right transversal of Ω_E in Ω_F .

Supplement to Theorem 10.

The diagram

$$\begin{array}{ccc} \mathrm{Hom}_{\Omega_F}(K_{A,F}, X) & \xrightarrow{\Lambda} & \mathrm{Hom}_{\Omega_F}(K_{B,F}, X) \\ \downarrow & & \downarrow \\ \mathrm{Hom}_{\Omega_E}(K_{A,E}, X) & \xrightarrow{N_{E/F}} & \mathrm{Hom}_{\Omega_F}(K_{A,E}, X) \end{array} \quad , \quad (X \text{ an } \Omega_F\text{-module}), \qquad (2.4)$$

commutes. Its columns (defined below) are bijective.

Here the left column comes from I.Propositions 2.1, 2.2, i.e., is induced by the inclusion $K_{A,E} \subset K_{A,F}$. The right column comes from I. Proposition 2.3 which implies an isomorphism $K_{A,E} \cong K_{B,F}$, being given by restricting representations T of A to B . One then uses this to transfer to $K_{A,E}$ the natural structure of $K_{B,F}$ as Ω_F-module. This in turn extends the natural structure of $K_{A,E}$ as Ω_E-module. We are in fact in exactly the same situation as that explained in the remarks of §1 leading up to diagram (1.5) - except that the symbols A and B have been interchanged.

Remark: The above was the form of the map on the Hom group used in [F7] (cf. Appendix A VI) in connection with restriction of basefield.

Proof of the supplement: We choose a basis $\{a_r\}$ of E over F , viewing it as a basis of A over B . If $a \in A$ then $\Lambda(a) = (\Lambda_{rs}(a))$ is given by the equation

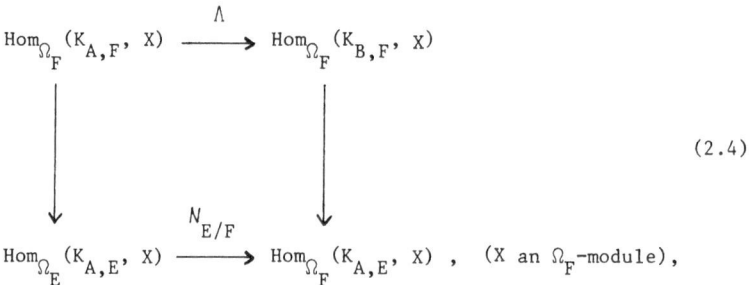

$$a\,a_s = \sum_r a_r\,\Lambda_{rs}(a), \qquad \Lambda_{rs}(a) \in B\ .$$

With $\{\sigma\}$ a right transversal of Ω_E in Ω_F we get equations

$$a^\sigma a_s^\sigma = \sum_r a_r^\sigma\ \Lambda_{rs}(a)$$

in $B \otimes_F F^c$, Ω_F acting via the right hand tensor factor. The matrix $\mathcal{A} = (a_s^\sigma)_{\sigma,s}$ is non-singular and thus we get a matrix equation

$$\mathcal{A}^{-1}\ \mathrm{diag}_\sigma(a^\sigma)\ \mathcal{A} = \Lambda(a) \tag{2.5}$$

where diag_σ is the diagonal matrix indexed by σ.

Let $T\colon B \to M_n(F_c)$ be a representation over F, with class χ in $K_{B,F}$. Then we obtain a representation $A \to M_n(F_c)$, which we shall denote by $T \otimes 1$ (abuse of notation), with class $\chi \otimes 1$. This is the compositum

$$A = B \otimes_F E \to M_n(F_c) \otimes_F E \to M_n(F_c),$$

$$x \otimes e \longmapsto xe\ .$$

The map $\chi \mapsto \chi \otimes 1$ is an isomorphism $K_{B,F} \cong K_{A,E}$. Its inverse underlies the isomorphism in the right hand column of (2.4), the latter being denoted for the moment by Θ. Thus, if in the sequel f is an element of $\mathrm{Hom}_{\Omega_F}(K_{A,F},\ \cdot)$ then

$$\Theta\ \Lambda f(\chi \otimes 1) = f(\Lambda'\chi)\ . \tag{2.6}$$

Here $\Lambda'\chi$ is the class in $K_{A,F}$ of $a \longmapsto T\Lambda(a)$. But by applying $T \otimes 1$ to (2.5), we see that the representation $T\Lambda$ of A over F is equivalent to the sum of representations T_σ, where for

$$a = \sum b_i \otimes e_i, \quad b_i \in B, \quad e_i \in E,$$

$$T_\sigma(a) = \sum T(b_i)\, e_i^\sigma .$$

T_σ is a representation $A \to M_n(F_c)$ over F. Note that

$$T_\sigma(a) = (\sum T(b_i)^{\sigma^{-1}}\, e_i)^\sigma .$$

Thus the class of T_σ is $(\chi^{\sigma^{-1}} \otimes 1)^\sigma \in K_{A,F}$. Here we first apply σ^{-1} to χ in the Ω_F-module $K_{B,F}$, then extend to $\chi^{\sigma^{-1}} \otimes 1 \in K_{A,E} \subset K_{A,F}$, then apply σ to this element in the Ω_F-module $K_{A,F}$. To recapitulate, we have $\Lambda'\chi = \sum_\sigma (\chi^{\sigma^{-1}} \otimes 1)^\sigma$. Note now firstly that, by (2.6),

$$\Theta\, \Lambda f(\chi \otimes 1) = \Pi\, f((\chi^{\sigma^{-1}} \otimes 1)^\sigma) . \tag{2.7}$$

Secondly we turn our attention to the left hand column of (2.4), and we denote by f' the image of f. Thus f' is just the restriction of f to $K_{A,E}$. But the structure of $K_{A,E}$ as Ω_F-module, as transferred from $K_{B,F}$, tells us that $\chi^\sigma \otimes 1$ is the image of $\chi \otimes 1$ under the action of σ. Hence indeed by (2.7)

$$\Theta\, \Lambda f(\chi \otimes 1) = N_{E/F}\, f'(\chi \otimes 1) ,$$

as we had to show.

Now we return to the general case.

Remark 1. The map $\Lambda'\colon K_{B,F} \to K_{A,F}$ coincides in all cases of the last theorem with the usual induction map. To see this let $T\colon B \to M_n(F_c)$ be a representation, making $V = F_c^n$ into a left $B \otimes_F F_c$-module, with T corresponding to an F_c-basis $\{v_i\}$ of V. Let $\{a_r\}$ be the B-basis of A_B, inducing $\Lambda\colon A \to M_m(B)$. Then in terms of the F_c-basis $\{a_r \otimes v_i\}$ of $A \otimes_B V$, we get for $a \in A$,

$$a(a_r \otimes v_i) = \sum_s a_s \Lambda_{sr}(a) \otimes v_i = \sum_s a_s \otimes \Lambda_{sr}(a)\, v_i$$

$$= \sum_{s,j} a_s \otimes v_j\, T_{ji}(\Lambda_{sr}(a)) ,$$

as we had to show.

Remark 2. Although the homomorphisms Λ commute with the maps of sequence I (1.1), the same is not true for I (1.2), except when A is actually free over B. Denote the class of a locally free module X in K_o by $[X]$ with the appropriate subscript and its image in the classgroup by (X). For a locally free A-module X of rank r we thus have to compare

$$\Lambda((X)_A) = \Lambda([X]_A - r[A]_A) = [X]_B - r[A]_B$$

with

$$(X)_B = [X]_B - rm[B]_B .$$

As $(A)_B = [A]_B - m[B]_B$, we get

2.1 Proposition:

$$(X)_{\mathcal{B}} = \Lambda((X)_{\mathcal{A}}) + r(A)_{\mathcal{B}} .$$

This generalises a classical formula (e.g. [F2]).

With $\mathcal{A}$, $\mathcal{B}$, A , B as before, we shall from now on assume that we have an involution $^-$ on A over F , with

$$\bar{B} = B, \quad \bar{\mathcal{A}} = \mathcal{A}, \quad \bar{\mathcal{B}} = \mathcal{B} .$$

We shall need one more element of structure namely a left and right B-linear map

$$t: {}_B A_B \to B \tag{2.8}$$

i.e., so that $t(b_1\ a\ b_2) = b_1\ t(a)\ b_2$, for all $b_1, b_2 \in B$, all $a \in A$, and so that furthermore

$$t(\bar{a}) = \overline{t(a)} . \tag{2.9}$$

t then defines a Hermitian form

$$\beta_{A/B} = \beta : A \times A \to B$$

by

$$\beta(a_1, a_2) = t(\bar{a}_1\ a_2), \quad \forall\ a_1, a_2 \in A . \tag{2.10}$$

We then require furthermore that β be non-singular.

Let again $\{a_r\}$ be a basis of A_B over B . Write

$$\mathcal{B} = (t(\bar{a}_r \, a_s)) \in H^o \, GL_m(B) \tag{2.11}$$

for the discriminant matrix of the form β . If $c = (c_{ij})_{i,j} \in GL_q(A)$ we define

$$\tilde{t}c = (t(\bar{a}_r \, c_{ij} \, a_s))_{(r,i),(s,j))} \, . \tag{2.12}$$

Thus $\mathcal{B} = \tilde{t}1$.

Examples: 1) $A = B \otimes_F E$, E a finite (separable) extension field of F , t the trace $E \to F$ extended to $A \to B$.

2) $\mathcal{A} = o\Gamma$, $\mathcal{B} = o\Delta$ with Δ a subgroup of the finite group Γ . Then (cf. [FMc1], [FMc2]) the right choice for t is the map which on Γ is given by

$$t(\gamma) = \begin{cases} 0 & \text{if } \gamma \in \Gamma \quad \gamma \notin \Delta \\ \gamma & \text{if } \gamma \in \Delta \, . \end{cases} \tag{2.13}$$

In this case $\mathcal{B}$ is the identity matrix, and β defines a non-singular form $\mathcal{A} \times \mathcal{A} \to \mathcal{B}$.

Theorem 11. (i) The maps Λ in Theorem 10 all commute with $^{-}$ action, with the possible exception of the map $K_o(\mathcal{A}) \to K_o(\mathcal{B})$. Here we have for $x \in K_o(\mathcal{A})$,

$$\Lambda\bar{x} - \overline{\Lambda x} = (r_{\mathcal{A}}x) \, ([\mathrm{Hom}_{\mathcal{B}}(\mathcal{A},\mathcal{B})]_{\mathcal{B}} - [\mathcal{A}]_{\mathcal{B}}) \, ,$$

with $r_{\mathcal{A}}$ *the rank map* $K_0(\mathcal{A}) \to \mathbb{Z}$.

If in particular β *defines a non-singular form* $\mathcal{A} \times \mathcal{A} \to \mathcal{B}$ *then* $\overline{\Lambda x} = \Lambda \bar{x}$.

(ii) *The map* $\Lambda' \colon K_{B,F} \to K_{A,F}$ *gives rise to a map* $\Lambda' \colon K^{s}_{B,F} \to K^{s}_{A,F}$ *and hence to maps*

$$\Lambda \colon \mathrm{Hom}_{\Omega_F}(K^{s}_{A,F}, \cdot) \to \mathrm{Hom}_{\Omega_F}(K^{s}_{B,F}, \cdot) .$$

These in turn yield maps

$$\Lambda \colon \mathrm{HCl}(\mathcal{A}) \to \mathrm{HCl}(\mathcal{B}), \quad \mathrm{HCl}(A) \to \mathrm{HCl}(B) ,$$

and for $\mathcal{O}$ *global*

$$\mathrm{Ad}\ \mathrm{HCl}(\mathcal{A}) \to \mathrm{Ad}\ \mathrm{HCl}(\mathcal{B}), \quad \mathrm{Ad}\ \mathrm{HCl}(A) \to \mathrm{Ad}\ \mathrm{HCl}(B) ,$$

all commuting with the maps P . *Moreover* Λ *commutes with the homomorphism* ν *of* II. *Proposition* 6.2. *and yields homomorphisms of the diagrams and sequences of Theorem* 4 (ii) *and of* II. *Proposition* 6.3 (ii).

(iii) *If* $(X_{\mathcal{A}}, h)$ *is a Hermitian* $\mathcal{A}$*-module then* $(X_{\mathcal{B}}, th)$ *is a Hermitian* $\mathcal{B}$*-module, where* $th(v,w) = t(h(v,w))$.

There result homomorphisms $\Lambda \colon K_0H(\mathcal{A}) \to K_0H(\mathcal{B})$, *and* $\widetilde{K_0H}(\mathcal{A}) \to \widetilde{K_0H}(\mathcal{B})$, *and a homomorphism of sequences* II (5.16), *where again* $\mathbb{Z} \to \mathbb{Z}$ *is multiplication by* m . *Analogously for* A, B *in place of* $\mathcal{A}$, $\mathcal{B}$.

For all $c \in GL_q(A)$, *we have* $\tilde{t}c \in GL_{mq}(B)$ (cf. (2.12)) *and*

$$\mathrm{Det}(\tilde{t}c) = \mathrm{Det}(\mathcal{B})^{q} \cdot \Lambda\,\mathrm{Det}(c) \,, \tag{2.14a}$$

i.e. for $\chi \in K_{B,F}$,

$$\mathrm{Det}_{\chi}(\tilde{t}c) = \mathrm{Det}_{\chi}(\mathcal{B})^{q} \cdot \mathrm{Det}_{\Lambda'\chi}(c) \,.$$

For $c \in H^{o}\,GL_{q}(A)$, *we have* $\tilde{t}c \in H^{o}(GL_{mq}(B))$ *and*

$$\mathrm{Pf}(\tilde{t}c) = \mathrm{Pf}(\mathcal{B})^{q} \cdot \Lambda\mathrm{Pf}(c) \,, \tag{2.14b}$$

i.e. for $\chi \in K^{s}_{B,F}$

$$\mathrm{Pf}_{\chi}(\tilde{t}c) = \mathrm{Pf}_{\chi}(\mathcal{B})^{q}\ \mathrm{Pf}_{\Lambda'\chi}(c) \,.$$

If (X_A, h) *is a Hermitian* A-*module then*

$$d_{\mathcal{B}}((X_{\mathcal{B}}, th)) = \Lambda d_{A}\,(X_A, h) \cdot d_{\mathcal{B}}((A_{\mathcal{B}}, \beta))^{r_A(X)}$$

and analogously for Hermitian A-*modules*.

In particular we get a commutative diagram

$$\begin{array}{ccc} \widetilde{K_o H}(A) & \xrightarrow{\ \Lambda\ } & \widetilde{K_o H}(\mathcal{B}) \\ \Big\downarrow d_A & & \Big\downarrow d_{\mathcal{B}} \\ \mathrm{HCl}(A) & \xrightarrow{\ \Lambda\ } & \mathrm{HCl}(\mathcal{B}) \,. \end{array}$$

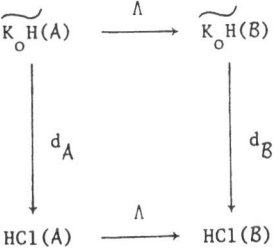

Proof: We first establish a formula for the behaviour of the map $\Lambda: A \to M_m(B)$ with respect to the involution, given in terms of a basis $\{a_r\}$ of A_B over B. For $a \in A$ we have

$$\sum_k t(\bar{a}_r a_k)\Lambda_{k,s}(\bar{a}) = \sum_k t(\bar{a}_r a_k \Lambda_{k,s}(\bar{a}))$$

$$= t(\bar{a}_r \bar{a} a_s) = t(\overline{aa_r}\, a_s) = \sum_j t(\overline{a_j \Lambda_{jr}(a)}\, a_s)$$

$$= \sum_j \overline{\Lambda_{jr}(a)}\, t(\bar{a}_j a_s) .$$

This gives the matrix equation

$$\Lambda(a)\mathcal{B} = \mathcal{B}\, \Lambda(\bar{a}) . \tag{2.15}$$

(Analogously for $M_n(A)$ in place of A).

Next we consider the map $T \mapsto \bar{T}$, for representations $T: B \to M_n(F_c)$ (cf. II (1.2)), and their extensions to

$$M_q(B) = M_q(F) \otimes_F B \to M_q(F) \otimes_F M_n(F_c) = M_{qn}(F_c) .$$

To avoid possible confusion here, we shall write, for the moment only, $1 \otimes T$ for such an extension. We then wish to show that

$$1 \otimes \bar{T} = \overline{1 \otimes T} , \tag{2.16}$$

where of course the right hand side is defined in terms of the standard matrix extensions of involutions (cf. II (1.1)). Let $b \in B$ and let e_{ij} be the matrix in $M_q(F)$ with 1 in place (i,j) and 0 elsewhere. Then

$$\overline{(1 \otimes T)}\,(e_{ij} \otimes b) = (1 \otimes T)_t\,\overline{(e_{ij} \otimes b)}$$

$$= (1 \otimes T)_t\,(e_{ji} \otimes \bar{b}) = e_{ij} \otimes T_t(\bar{b})$$

$$= e_{ij} \otimes \bar{T}(b) = (1 \otimes \bar{T})\,(e_{ij} \otimes b)\ ,$$

as required.

In the notation of (2.16), and by (2.15), we now have

$$\Lambda'\bar{T}(a) = (1 \otimes \bar{T})\,(\Lambda(a)) = \overline{(1 \otimes T)}\,(\Lambda(a))$$

$$= ((1 \otimes T)\,\overline{(\Lambda(a))})_t = ((1 \otimes T)\,(\mathcal{B}\Lambda(\bar{a})\mathcal{B}^{-1}))_t$$

$$= ((1 \otimes T)\,(\mathcal{B}))_t^{-1}\ ((1 \otimes T)\,(\Lambda(\bar{a})))_t\ ((1 \otimes T)\,(\mathcal{B}))_t$$

$$= ((1 \otimes T)\,(\mathcal{B}))_t^{-1} \cdot \overline{\Lambda'T}(a) \cdot ((1 \otimes T)\,(\mathcal{B}))_t\ .$$

Thus $\Lambda'\bar{T}$ and $\overline{\Lambda'T}$ are equivalent, i.e., the map $\Lambda'\colon K_{B,F} \to K_{A,F}$ preserves $^-$ action, hence so does $\Lambda\colon \mathrm{Hom}_{\Omega_F}(K_{A,F},\cdot) \to \mathrm{Hom}_{\Omega_F}(K_{B,F}\,\cdot)$. Equation (2.15) also implies that the map $\Lambda\colon K_1A \to K_1B$ preserves $^-$ action. For $K_1\mathcal{A} \to K_1\mathcal{B}$ one needs a slightly more sophisticated argument. In principle it is the same and we shall not give it here, as the result is not needed elsewhere. The remainder of (i) - except for the formula concerning $K_o(\mathcal{A}) \to K_o(\mathcal{B})$ - now follows easily. Thus e.g. consider in the local case the commutative diagram with exact rows,

$$\begin{array}{ccccccccc}
1 \to & \mathrm{Det}\,\mathcal{A}^* & \to & \mathrm{Hom}_{\Omega_F}(K_{A,F},\, F_c^*) & \to & K_oT(\mathcal{A}) & \to 1 \\
 & \downarrow & & \downarrow & & \downarrow & \\
1 \to & \mathrm{Det}\,\mathcal{B}^* & \to & \mathrm{Hom}_{\Omega_F}(K_{B,F},\, F_c^*) & \to & K_oT(\mathcal{B}) & \to 1 .
\end{array}$$

As the middle column preserves $^-$ action so do the other two columns. From the local result we then get the global result for $K_oT(\mathcal{A}) \to K_oT(\mathcal{B})$. Analogous arguments work for $\mathrm{Det}\,A^* \to \mathrm{Det}\,B^*$ etc.

Now let X be a locally free $\mathcal{A}$-module spanning a free A-module V of rank $q = r_{\mathcal{A}} X$. The dual module $U = \mathrm{Hom}_A(V,A)$ is defined via a non-singular pairing

$$< , > : U \times V \to A .$$

Writing $[u,v] = t(<u,v>)$ we get a pairing $[,]$, which identifies U with $\mathrm{Hom}_B(V,B)$. Thus U is spanned by each of the locally free $\mathcal{B}$-modules

$$\mathrm{Hom}_{\mathcal{A}}(X,\mathcal{A}) = [u \in U \mid <u,X> \subset \mathcal{A}] ,$$

$$\mathrm{Hom}_{\mathcal{B}}(X,\mathcal{B}) = [u \in U \mid [u,X] \subset \mathcal{B}] .$$

We shall then show that, in $K_oT(\mathcal{B})$,

$$[\mathrm{Hom}_{\mathcal{B}}(X,\mathcal{B})/\mathrm{Hom}_{\mathcal{A}}(X,\mathcal{A})] = q[\hat{\mathcal{A}}_{\mathcal{B}}/\mathcal{A}_{\mathcal{B}}] . \tag{2.17}$$

Here we write for short $\hat{\mathcal{A}}_{\mathcal{B}} = \mathrm{Hom}_{\mathcal{B}}(\mathcal{A},\mathcal{B})$, viewed as the submodule of A consisting of the elements y with $t(\bar{y}\,\mathcal{A}) \subset \mathcal{B}$ - in other words in the special case $V = A$ we take $< , >$ as the product. Moreover we have made the convention that for locally free $\mathcal{B}$-modules Y_1, Y_2 spanning the same B-module, $[Y_1/Y_2] = [Y_1/Y_1 \cap Y_2] - [Y_2/Y_1 \cap Y_2]$. Assuming (2.17) we get, on applying $\delta\colon K_oT(\mathcal{B}) \to K_o(\mathcal{B})$, and recalling II (1.8),

$$\overline{\Lambda([X]_A)} - \Lambda(\overline{[X]_A}) = - [\mathrm{Hom}_{\mathcal{B}}(X,\mathcal{B})]_{\mathcal{B}} + [\mathrm{Hom}_A(X,A)]_{\mathcal{B}}$$

$$= \delta([\mathrm{Hom}_{\mathcal{B}}(X,\mathcal{B})/\mathrm{Hom}_A(X,A)]_{\mathcal{B}}) = q\delta([\hat{A}_{\mathcal{B}}/A_{\mathcal{B}}])$$

$$= q([A]_{\mathcal{B}} - [\mathrm{Hom}_{\mathcal{B}}(A,\mathcal{B})]_{\mathcal{B}}) ,$$

as we had to show.

For the proof of (2.17) we may suppose $\mathcal{O}$ to be local. Take $\{x_j\}$ as A-basis of X, $\{y_i\}$ as the dual A-basis of $\mathrm{Hom}_A(X,A)$. Thus $\langle y_j, x_i\rangle = \delta_{ji}$ (Kronecker-symbol). Let $\{a_r\}$ be a $\mathcal{B}$-basis of $A_{\mathcal{B}}$, $\{c_r\}$ the $\mathcal{B}$-basis of $\hat{A}_{\mathcal{B}}$, so that $t(\bar{c}_r\, a_s) = \delta_{rs}$. Then $\{y_j\, a_r\}$ is a $\mathcal{B}$-basis of $\mathrm{Hom}_A(X,A)$, $\{y_j\, c_r\}$ one of $\mathrm{Hom}_{\mathcal{B}}(X,\mathcal{B})$. If now $f \in GL_m(B)$ transforms the B-basis $\{c_r\}$ of A into the B-basis $\{a_r\}$, then $\mathrm{diag}(f) \in GL_{mq}(B)$ transforms the B-basis $\{y_j\, c_r\}$ of $U = \mathrm{Hom}_A(V,A)$ into the B-basis $\{y_j\, a_r\}$. As $\det(\mathrm{diag}(f)) = \det(f)^q$, (2.17) follows, using the description $K_oT(\mathcal{B}) = \mathrm{Det}\, B^*/\mathrm{Det}\, \mathcal{B}^*$.

<u>Remark</u>: If actually $\hat{A}_{\mathcal{B}} \supset A_{\mathcal{B}}$ then we have shown that $\mathrm{Hom}_{\mathcal{B}}(X,\mathcal{B})/\mathrm{Hom}_A(X,A)$ is a module isomorphic to the direct sum of q copies of $\hat{A}_{\mathcal{B}}/A_{\mathcal{B}}$.

We have now completed the proof of (i). For (ii) let $T : B \to M_n(F_c)$ be a representation, with $T(\bar{b}) = T(b)^j$, j a symplectic involution of $M_n(F_c)$. Denote by j also its matrix extension to $M_m(M_n(F_c)) = M_{mn}(F_c)$. Define a new involution k on $M_{mn}(F_c)$ by

$$P^k = T(\mathcal{B})^{-1}\, P^j\, T(\mathcal{B}) , \tag{2.18}$$

with $\mathcal{B}$ as in (2.11). Here we may again use the simpler notation $T(\mathcal{B})$ in place of $1 \otimes T(\mathcal{B})$. k is symplectic, as $T(\mathcal{B})^j = T(\bar{\mathcal{B}}) = T(\mathcal{B})$. By (2.15), $\Lambda' T(\bar{a}) = \Lambda' T(a)^k$. Thus $\Lambda' T$ is symplectic, if T is. The remainder of (ii) is now easily derived. For instance we now know that Λ maps $\mathrm{Hom}_{\Omega_F}(K^s_{A,F}, \cdot)$ into $\mathrm{Hom}_{\Omega_F}(K^s_{B,F}, \cdot)$, and by restriction from Λ: Det $A^* \to$ Det B^* we get Λ: $\mathrm{Det}^s A^* \to \mathrm{Det}^s B^*$. Hence we get an induced homomorphism Λ: $\mathrm{HCl}(A) \to \mathrm{HCl}(B)$. Similarly in the other cases.

The first part of (iii) is again obvious, including the existence of homomorphisms on $K_o H$. Next we shall prove (2.14b). The proof of (2.14a) is similar, but easier. With $\tilde{t}c$ given as in (2.12), we get

$$(\tilde{t}c)_{(r,i)(s,j)} = \sum_v t(\bar{a}_r a_v) \Lambda_{vs}(c_{ij}),$$

whence with appropriate arrangement of indices

$$\tilde{t}c = \begin{pmatrix} \mathcal{B} & & & \\ & \ddots & & \\ & & \ddots & \\ & & & \mathcal{B} \end{pmatrix} \Lambda(c) = \mathrm{diag}(\mathcal{B}) \cdot \Lambda(c),$$

where the first matrix on the right hand side is diagonal with q blocks. Now let T be a symplectic representation of B, $T(\bar{b}) = T(b)^j$, j the appropriate symplectic involution of $GL_n(F_c)$. We now apply II. Proposition 3.6 with $R = T(\mathrm{diag}(\mathcal{B}))$, $S = T(\Lambda(c))$. Clearly R and RS are j-symmetric. We get

$$\mathrm{Pf}_\chi(\tilde{t}c) = \mathrm{Pf}^j(T(\tilde{t}c)) = \mathrm{Pf}^j(T(\mathrm{diag}(\mathcal{B}))) \, \mathrm{Pf}^k(T\Lambda(c))$$

$$= \mathrm{Pf}_\chi(\mathcal{B})^q \, \mathrm{Pf}_{\Lambda'\chi}(c),$$

as required. We have of course to justify here the equation $Pf^k(T\Lambda(c)) = Pf_{\Lambda'\chi}(c)$. By (2.15) and (2.18) we know that $\Lambda'T$ takes $^-$ into k, hence the matrix extension of $^-$ into that of k. But one verifies easily that if conjugation by $T(\mathcal{B})$ takes j into k then conjugation by $\mathrm{diag}(T(\mathcal{B})) = T(\mathrm{diag}(\mathcal{B}))$ takes the matrix extension of j into that of k. This then justifies the above equation.

Finally we come to the discriminant formula. We shall prove this assuming o to be global. The local case and that for $o = F$ are similar, but easier. Let then (X_A, h) be a Hermitian A-module, with X_A spanning an A-module V of rank q. Let $\{v_i\}$ be a basis of V over A, and let Y be the free A-module on $\{v_i\}$. Then $d((X_A, h))$ is represented by

$$(\mathrm{Det}\,\alpha,\ \mathrm{Pf}(h(v_i, v_j))) \in \mathrm{Det}\, JA \times \mathrm{Hom}_{\Omega_F}(K^s_{A,F}, F^*_c)$$

where, for each p, $\alpha_p Y_p = X_p$. Next we now take $\{v_i a_r\}$ as a basis of V over B, with $\{a_r\}$ a basis of A_B over B. For given p, let $\{x_{ip}\}$ be a basis of X_p over A_p and $\{c_{rp}\}$ a basis of A_p over B_p. Then the matrix $\alpha_p \in GL_q(A_p)$, considered above, may be viewed as taking $\{v_i\}$ into $\{x_{ip}\}$. Moreover $\Lambda(\alpha_p)$ will take $\{v_i a_r\}$ into $\{x_{ip} a_r\}$. If now $\gamma_p \in GL_m(B_p)$ takes $\{a_r\}$ into $\{c_{rp}\}$ then, with appropriate ordering of bases, the diagonal matrix of q blocks, $\mathrm{diag}(\gamma_p)$, will take $\{x_{ip} a_r\}$ into $\{x_{ip} c_{rp}\}$. Therefore $d((X_B, th))$ is represented by

$$(\mathrm{Det}\,\Lambda(\alpha)\,\mathrm{Det}(\gamma)^q,\quad \mathrm{Pf}(\Lambda(h(v_i, v_j)))\,\mathrm{Pf}(\mathcal{B})^q),$$

where for the second entry we have used (2.14). As $(\mathrm{Det}(\gamma), \mathrm{Pf}(\mathcal{B}))$ represents (A_B, β) the required equation now follows.

Remark 1. The maps Λ will in general not commute with those of sequence II (5.17).

Remark 2. Both Proposition 2.1 and the discriminant formula in the last theorem suggest that one should look at the map

$$[(X_A,h)] \mapsto [(X_B,th)] - (r_A X)[(A_B,\beta)] .$$

Remark 3. The map $K^s_{B,F} \to K^s_{A,F}$ coincides again with induction for symplectic representations (e.g. defined in [FMc2]).

Remark 4. Analogously to the involution free case (cf. addendum to Theorem 10), and in the same situation, i.e., with $A = B \otimes_F E$, and t given by the trace $E \to F$, we get

Supplement to Theorem 11. The maps of diagram (2.4) preserve $^-$ action and give rise to a commutative diagram with bijective columns

$$\begin{array}{ccc} \mathrm{Hom}_{\Omega_F}(K^s_{A,F}, X) & \xrightarrow{\ \Lambda\ } & \mathrm{Hom}_{\Omega_F}(K^s_{B,F}, X) \\ \Big\downarrow \wr & & \Big\downarrow \wr \\ \mathrm{Hom}_{\Omega_E}(K^s_{A,E}, X) & \xrightarrow{\ N_{E/F}\ } & \mathrm{Hom}_{\Omega_F}(K^s_{A,E}, X) . \end{array} \qquad (2.19)$$

Proof: Action by Galois groups preserves $^-$ action and the symplectic property. Hence $N_{E/F}$ preserves $^-$ action and the symplectic property. The same has already been established for Λ. The isomorphism $K_{A,E} \cong K_{B,F}$, underlying the right hand column of (2.4) preserves $^-$ action and yields an isomorphism $K^s_{A,E} \cong K^s_{B,F}$. This gives rise to the right hand column of diagram (2.19). To come to the left hand column, the embedding $K_{A,E} \subset K_{A,F}$ again clearly preserves $^-$ action and the symplectic property; to show that the left hand column of (2.19) is actually bijective one uses the fact that $K^s_{A,F}$ is closed under Ω_F-action. Commutativity follows in the same way as that of (2.4).

CHAPTER V. GROUPS

Here we shall apply the general theory, developed so far, to group rings and add further results valid specifically for these. The involution on the group ring $F\Gamma$ of a finite group Γ over a field F is always the F-linear map $F\Gamma \to F\Gamma$ which takes every group element γ into γ^{-1}. We shall give briefly a translation of the language of representations of algebras to that of representations of groups and of group characters, both in the linear context where it is well known, as well as in the Hermitian one. The principal new property which makes its appearance is that of virtual characters to form a ring rather than just an additive group, and of certain rings of characters to act on classgroups and on Hom groups in a consistent manner. This is also true in the Hermitian theory. In principle one could consider more general co-algebras rather than group rings, but we shall not do so here. Our theorems, in IV, on change of orders are now applied in the context both of change of group, and of change of basefield, and then lead to structures of Frobenius modules.

For the basic facts in the theory of group representations see [Se1] and [CR] .

§1. Characters

We shall keep to the over-all assumption that fields have characteristic zero - except where otherwise stated. (The results of this section will of course extend to some other characteristics).

Let Γ be a finite group, L a field. We consider representations, i.e. group homomorphisms $T: \Gamma \to GL_n(L)$, for varying degree n . With the usual notion of equivalence, and of sum

$$T_1, T_2 \longmapsto \begin{pmatrix} T_1 & 0 \\ 0 & T_2 \end{pmatrix}$$

of representations, we get an Abelian semigroup, whose Grothendieck group is denoted by $R_{\Gamma,L}$, or just by R_Γ if L is algebraically closed and there is no danger of confusion. Associate with each representation T its character χ, viewed as a function $\Gamma \to L$ and defined by $\chi(\gamma) = \text{trace } T(\gamma)$. The equivalence classes of representations correspond biuniquely to the characters. R_Γ can and will thus be viewed as the additive group of "virtual characters" under pointwise addition: $(\chi+\chi')(\gamma) = \chi(\gamma) + \chi'(\gamma)$. Pointwise multiplication of characters, or equivalently the tensor or "Kronecker" product of representations makes $R_{\Gamma,L}$ into a commutative ring (all rings have an identity!)

Next let F be a field, and $L = F_c$ its algebraic closure. Then the Galois group Ω_F acts on the ring $R_\Gamma = R_{\Gamma,L}$. In terms of representations we have $T^\omega(\gamma) = T(\gamma)^\omega$, for $\omega \in \Omega_F$, in terms of characters $\chi^\omega(\gamma) = \chi(\gamma)^\omega$. The subring $R_{\Gamma,F}$ is fixed under Ω_F, but not necessarily equal to the fixed ring $R_\Gamma^{\Omega_F}$.

Given a representation $T: \Gamma \to GL_n(L)$ we obtain a representation $\bar{T}$, where

$$\bar{T}(\gamma) = T(\gamma^{-1})_t, \quad (X_t \text{ the transpose of } X). \tag{1.1}$$

This defines an involutory automorphism $\chi \longmapsto \bar{\chi}$ of the ring $R_{\Gamma,L}$. Next let β be a non-singular symmetric (skew symmetric) bilinear form on a finite dimensional L-vector space V. A representation T of Γ which is the compositum

$$\Gamma \to \text{Aut}(\beta) \to \text{Aut}_L(V) = GL_n(L),$$

will be called an <u>orthogonal</u> (<u>symplectic</u>) representation of Γ. Going over to characters we get the subgroups R_Γ^o *(orthogonal virtual characters)* and R_Γ^s *(symplectic virtual characters)* of R_Γ (taking L algebraically closed).

Remark: Strictly speaking one should reserve e.g. the term orthogonal representation to the homomorphisms $\Gamma \to \mathrm{Aut}(\beta)$, β symmetric. As, however, L is algebraically closed the transition from these to the linear representations $\Gamma \to \mathrm{Aut}_L(V)$ is injective on equivalence classes, i.e. yields an embedding of Grothendieck groups. Analogously in the symplectic case.

1.1 *Proposition:* *Extension of representations* $\Gamma \to GL_n(F_c)$ *of groups to representations* $F\Gamma \to M_n(F_c)$ *of* F-*algebras yields an isomorphism* $R_\Gamma \cong K_{F\Gamma,F}$ *of* Ω_F-*modules and* ¯ *modules, and similarly an isomorphism* $R_\Gamma^s \cong K_{F\Gamma,F}^s$ *(and likewise for* R_Γ^o) . *Moreover if* $F_c \supset E \supset F$ *then the diagram*

$$
\begin{array}{ccc}
 & & K_{F\Gamma,F} \\
 & \nearrow & \downarrow \\
R_\Gamma & & \\
 & \searrow & \downarrow \\
 & & K_{E\Gamma,E}
\end{array}
\tag{1.2}
$$

commutes, the column coming from I *Proposition* 2.3.

In the situation considered in I §2 and in II we may now view the map Det as going into the Hom groups $\mathrm{Hom}_{\Omega_F}(R_\Gamma, \cdot)$ or $\mathrm{Hom}_{\Omega_F}(R_\Gamma^s, \cdot)$. Then we have

Corollary to Proposition 1.1 *In the isomorphisms of Theorem* 2, *and in the definition of* HCl *and* Ad HCl, *and the various Theorems and Propositions in* II, *the groups* $K_{F\Gamma,F}$ *and* $K_{F\Gamma,F}^s$ *can always be replaced by* R_Γ *and* R_Γ^s *respectively, provided that* $A = F\Gamma$, *and*, where appropriate, $A = o\Gamma$.

We shall next restate the definitions of Frobenius functor and Frobenius module (cf. [L], [Ba2]) in a manner convenient to us, and we shall also introduce an intermediary concept which will be found useful. Let C be a category, C_o a subcategory with the same objects as C . A Frobenius functor F of C , C_o is a pair F^* , F_* of functors, where F^* is contravariant from C to commutative rings (always with 1), F_* is covariant from C_o to Abelian groups, and where on objects X in C_o, F^* and F_* coincide - more precisely where $F_*(X)$ is the additive group of $F^*(X)$. (The use of the symbol $*$, as in F^* , here and in similar contexts in the sequel, should not give rise to confusion with its general use elsewhere to indicate groups of invertible elements). Moreover if $\alpha: X \to Y$ is a morphism in C_o and if we write for simplicity

$$F^*(\alpha) = \alpha^*, \quad F_*(\alpha) = \alpha_* , \tag{1.3}$$

the identity

$$\alpha_*(x \cdot \alpha^* y) = (\alpha_* x) \cdot y , \qquad x \in F(X) , \qquad y \in F(Y) \tag{1.4}$$

is to be satisfied. In the usual definition $C = C_o$. The new definition is more convenient when we come to Frobenius modules. A covariant (contravariant) pre-Frobenius module M over a Frobenius functor F is given by (i) a pair M^*, M_* of functors into Abelian groups, where again $M^*(X) = M_*(X) = M(X)$ on objects X , and (ii) on $M(X)$ the structure of an $F(X)$-module (for all objects X in C) . M is covariant, if M^*, M_* have the same domain and variance as F^*, F_* respectively, i.e., M^* is a contravariant functor of C, M_* a covariant functor of C_o . M is contravariant if M^* is a contravariant functor of C_o, M_* a covariant functor of C . A pre-Frobenius module M over F (of either variance) is a Frobenius module over F , if the following conditions are satisfied. Let

$\alpha: X \to Y$ be a morphism in $\mathcal{C}$, use the notation (1.3) and in addition also write (whenever the left hand sides are defined)

$$M^*(\alpha) = \alpha^* , \quad M_*(\alpha) = \alpha_* . \tag{1.3a}$$

Then for all $x \in \mathcal{F}(X)$, $y \in \mathcal{F}(Y)$, $u \in M(X)$, $v \in M(Y)$, we must have

$$\alpha^*(y \cdot v) = \alpha^*(y)\ \alpha^*(v) , \tag{1.5}$$

$$\alpha_*(x \cdot \alpha^* v) = (\alpha_* x) \cdot v , \tag{1.6}$$

$$\alpha_*(\alpha^* y \cdot u) = y \cdot \alpha_* u . \tag{1.7}$$

Note that if M is covariant we need $\alpha \in \mathcal{C}_o$ for (1.6) and (1.7), and if M is contravariant we need $\alpha \in \mathcal{C}_o$ for (1.5) and (1.6).

<u>Remark 1.</u> We do not need the (obvious) notion of a pre-Frobenius functor.

<u>Remark 2.</u> Suppose that M_*, M^* are functors into Ω-modules, Ω a group and that the action of $\mathcal{F}(X)$ on $M(X)$ commutes with the action of Ω. Then all our definitions extend and we shall speak of Frobenius modules <u>with</u> Ω-<u>module values</u>.

We now come to the Lemma which is the principal reason for introducing pre-Frobenius modules as a useful device. The notions of a sub or quotient pre-Frobenius module, and of a product of pre-Frobenius modules, - always over a fixed Frobenius functor and of the same variance - are the obvious ones.

1.2 Lemma: Let F be a Frobenius functor.

(i) If F is a Frobenius subfunctor of F', then F' is a covariant Frobenius module over F.

(ii) Let Ω be a group, G an Ω-module, M a Frobenius module over F with Ω-module values. Then $M'(X) = \mathrm{Hom}_{\Omega}(M(X), G)$ defines a Frobenius module over F of the opposite variance to M.

(iii) Any sub pre-Frobenius module and any quotient pre-Frobenius module of a Frobenius module over F is a Frobenius module over F of the same variance.

(iv) Any product of Frobenius modules over F of given variance is a Frobenius module of the same variance.

Proof: Obvious, although (ii) requires some computations.

Remarks on notation: By abuse of notation we shall in the sequel often speak of a "Frobenius functor $F(X)$" - X to be considered as a variable in this case.

Our first example is the category $C = G$ of finite groups (with appropriate restrictions on the orders or indices when fields of arbitrary characteristic are considered). The sub-category $C_o = G_o$ is that of embeddings. In the notation previously introduced $R_{\Gamma,F}$ and in particular $R_{\Gamma} = R_{\Gamma,F_c}$ are Frobenius functors of G, G_o. We shall use the customary notation. If $\Delta \subset \Gamma$, then we have the restriction map

$$\mathrm{res}^{\Gamma}_{\Delta}: R_{\Gamma} \to R_{\Delta} , \tag{1.8}$$

and the induction

$$\mathrm{ind}_{\Gamma}^{\Delta}: R_{\Delta} \to R_{\Gamma} , \tag{1.9}$$

and if Γ is a quotient group of Δ , the inflation map

$$\mathrm{inf}_{\Delta}^{\Gamma}: R_{\Gamma} \to R_{\Delta} . \tag{1.10}$$

These notations, as applied also to $R_{\Gamma,F}$ for any field F will be used in the sequel.

Let next Ω be a group which acts on the R_{Γ} , for variable Γ , in such a way that the maps (1.8) - (1.10) preserve this action. Dually to these maps we obtain homomorphisms

$$\mathrm{ind}_{\Gamma}^{\Delta}: \mathrm{Hom}_{\Omega}(R_{\Delta},G) \to \mathrm{Hom}_{\Omega}(R_{\Gamma},G) , \tag{1.8a}$$

$$\mathrm{res}_{\Delta}^{\Gamma}: \mathrm{Hom}_{\Omega}(R_{\Gamma},G) \to \mathrm{Hom}_{\Omega}(R_{\Delta},G) , \tag{1.9a}$$

$$\mathrm{coinf}_{\Gamma}^{\Delta}: \mathrm{Hom}_{\Omega}(R_{\Delta},G) \to \mathrm{Hom}_{\Omega}(R_{\Gamma},G) , \tag{1.10a}$$

of functors of Ω-modules G .

The second example is the category $\mathcal{C} = \mathcal{K}$, the opposite of the category of fields (of characteristic zero) and embeddings. Here the morphisms of $\mathcal{C}_o$ are the embeddings $F \subset E$, with the degree $[E:F]$ finite. Then $R_{\Gamma,F}$, as a functor of F , is a Frobenius functor. Thus for $E, F \in \mathcal{K}, E \supset F$, we have the ring embedding

$$\varepsilon_{E}^{F}: R_{\Gamma,F} \to R_{\Gamma,E} , \tag{1.11}$$

and, for $[E:F]$ finite, the'restriction of scalar map' of Abelian groups

$$\rho_F^E: R_{\Gamma,E} \to R_{\Gamma,F} , \tag{1.12}$$

under which E-linear automorphisms are viewed as F-linear ones. In terms of the characters χ we obtain

$$(\rho_F^E \chi)(\gamma) = t_{E/F}(\chi(\gamma)) , \tag{1.13}$$

with $t_{E/F}$ the trace.

In the above examples we really have functors in two "commuting" variables Γ, F. We shall then refer to these as <u>Frobenius bi-functors</u> and <u>Frobenius bi-modules</u>.

The preceding remarks (and notations) also apply to R_Γ^o (orthogonal representations), to $R_{\Gamma,F}^o = R_\Gamma^o \cap R_{\Gamma,F}$ and to R_Γ^s (symplectic representations). More precisely R_Γ^o and $R_{\Gamma,F}^o$ are Frobenius functors, the latter with respect to both indicated variables. Any Frobenius module e.g. over $R_{\Gamma,F}$ will, by restriction of scalars be a Frobenius module over $R_{\Gamma,F}^o$ - similarly for R_Γ and R_Γ^o. Moreover R_Γ^s is a Frobenius module over R_Γ^o and over $R_{\Gamma,F}^o$.

In view of the arithmetic applications, it is important to have strong induction theorems. These should enable one to reduce proofs to extensions with special Galois groups. To do this, we consider representations (say in characteristic zero)

$$T: \Gamma \to GL_2(F_c) ,$$

and we introduce a terminology for T in terms of generators of $\mathrm{Im}\, T$.

(a) T <u>bicyclic</u> $\longleftrightarrow$ $\mathrm{Im}\, T$ generated by $\begin{pmatrix} \eta & 0 \\ 0 & \eta^{-1} \end{pmatrix}$,

(b) T <u>dihedral</u> $\longleftrightarrow$ $\mathrm{Im}\, T$ generated by $\begin{pmatrix} \eta & 0 \\ 0 & \eta^{-1} \end{pmatrix}$, $\begin{pmatrix} 0 & 1 \\ 1 & 0 \end{pmatrix}$,

(c) T quaternion $\longleftrightarrow$ Im T generated by $\begin{pmatrix} \eta & 0 \\ 0 & \eta^{-1} \end{pmatrix}, \begin{pmatrix} 0 & 1 \\ -1 & 0 \end{pmatrix}$.

Here η is a root of unity, for (b) restricted to be of order $m > 2$, for (c) restricted to be of order $2m$ with $m \geq 2$. We extend this terminology to equivalent representations and to the associated characters. Clearly dihedral and quaternion representations of Γ are lifted from irreducible, faithful representations of dihedral or quaternion quotients, respectively, of Γ.(We use "quaternion" in the generalised sense). Evidently dihedral representations are orthogonal, quaternion ones are symplectic and bicyclic are both.

To state the main induction theorem we consider semidirect products

$$\left.\begin{array}{l} \Delta \rtimes \Pi\ ,\ \Pi \text{ a p-group, } \Delta \text{ a normal cyclic subgroup of} \\ \text{order prime to } p\ , \text{ with } \Pi \text{ acting on } \Delta \text{ via a} \\ \text{homomorphism } g\colon \Pi \to \pm 1\ , \text{ where} \\ \pi^{-1}\,\delta\,\pi = \delta^{g(\pi)}\ , \text{ all } \pi \in \Pi\ , \text{ all } \delta \in \Delta \end{array}\right\} \quad (1.14)$$

Theorem 12. Every orthogonal (resp. symplectic) character of Γ can be written in the form

$$\chi = \sum_i n_i \operatorname{ind}_{\Gamma}^{\Gamma_i} \chi_i \qquad (1.15)$$

where (i) the n_i are integers, (ii) Γ_i is a subgroup of Γ of type (1.14), (iii) for χ orthogonal, the χ_i are dihedral, or bicyclic, or Abelian (i.e. of degree 1) of order 1 or 2, and for χ symplectic, the χ_i are quaternion, or bicyclic.

This theorem is due to Serre. Its "orthogonal" part was published by him in [Se2]. It uses a deep result in [Bo-S]. Serre's complete proof was also published in [Mar]. Recently Ritter has given a more elementary proof (cf.[Ri]).

§2. Character action, ordinary theory

In the present section we connect (without considering at this stage Hermitian structure) the action of $R_{\Gamma,F}$ on classgroups and Grothendieck groups with that on Hom-groups. The former action occurs in Swan's work (cf.[SE]) (and see [L]) and in a different variant was considered by Chase (unpublished), the latter action was first observed by S. Ullom who derived the good behaviour of determinants (cf. [U]) (see also [Mt]). Going beyond this we shall also establish the Frobenius properties of these actions. Although in principle $\mathcal{O}$ may be taken as an arbitrary Dedekind domain, provided that the finite groups Γ have order prime to the characteristic of its quotient field F, we shall specifically think, and find it convenient to state our results, in terms of the three cases considered earlier. Accordingly it will throughout be assumed that all fields F to be considered are finite extensions of a fixed base field k (of characteristic zero), and that F is the quotient field of $\mathcal{O} = \mathcal{O}_F$ where either (i) $k = \mathbb{Q}_p$, $\mathcal{O}$ the ring of integers in F (local case), (ii) $k = \mathbb{Q}$, $\mathcal{O}$ the ring of algebraic integers in F (global case) (iii) k arbitrary, $\mathcal{O} = F$ ("field case"). In this last case parts of the next theorem become irrelevant. We shall also use the conventions of the Corollary to Proposition 1.1 without further reference. - We make $\mathbb{Z}$ into an R_Γ-module by $(\chi,n) \longmapsto \deg(\chi)n$, $\deg(\chi)$ the degree.

Theorem 13. (i) Let G be an Ω_k-module. The map

$$\chi, f \longmapsto \chi f \quad \text{where} \quad \chi f(\phi) = f(\bar{\chi}\phi)$$

with $\chi \in R_{\Gamma,F}$, $\phi \in R_\Gamma = R_{\Gamma,F_c}$, $f \in \mathrm{Hom}_{\Omega_F}(R_\Gamma, G)$, defines on $\mathrm{Hom}_{\Omega_F}(R_\Gamma, G)$ the structure of an $R_{\Gamma,F}$-Frobenius bi-module. More precisely, it is by duality to the Frobenius module R_Γ a contravariant Frobenius module over $R_{\Gamma,F}$, both as a functor of Γ (cf. (1.8) - (1.10)) and as a functor of F (cf. (1.11), (1.12)).

If $F \subset E$ the map $\text{Hom}_{\Omega_F}(R_\Gamma, G) \to \text{Hom}_{\Omega_E}(R_\Gamma, G)$ is just inclusion and the map $\text{Hom}_{\Omega_E}(R_\Gamma, G) \to \text{Hom}_{\Omega_F}(R_\Gamma, G)$ is the "norm map" $N_{E/F}$ (cf. IV (2.3)). Moreover homomorphisms $G \to G_1$ of Ω_k-modules induce homomorphisms of Frobenius modules.

(ii) $\text{Det}((F\Gamma)^*)$, and in the local case $\text{Det}((o\Gamma)^*)$ are Frobenius submodules of $\text{Hom}_{\Omega_F}(R_\Gamma, F_c^*)$; in the global case $\text{Det}(U(o\Gamma))$ and $\text{Det}(J(F\Gamma))$ are Frobenius submodules of $\text{Hom}_{\Omega_F}(R_\Gamma, J(F_c))$, always over $R_{\Gamma,F}$ and with respect to each of the variables Γ and F.

(iii) Let Z be a right $o\Gamma$-lattice, free over o. The functor $W \longmapsto W \otimes_o Z$ from $o\Gamma$-modules to $o\Gamma$-modules is exact from locally free to locally free $o\Gamma$-modules, and from l.f.p. torsion to l.f.p. torsion $o\Gamma$-modules. Both on $K_o(o\Gamma)$ and on $K_oT(o\Gamma)$, the map $[W] \longmapsto [W \otimes_o Z]$ is an endomorphism which only depends on the character χ_Z, associated with the representation of Γ on $Z \otimes_o F_c$. These characters generate $R_{\Gamma,F}$, and the actions of the modules Z define on $K_o(o\Gamma)$, on $\text{Cl}(o\Gamma)$ and on $K_oT(o\Gamma)$, respectively, the structure of $R_{\Gamma,F}$-modules. Moreover with respect to these actions and for varying Γ and varying F, the groups $K_oT(o\Gamma)$, $K_o(o\Gamma)$, $\text{Cl}(o\Gamma)$ and $\mathbb{Z}$ are contravariant Frobenius modules over $R_{\Gamma,F}$ and they are Frobenius bi-modules. If $\Delta \to \Gamma$ is a homomorphism of groups the corresponding covariant maps are given by $[W] \longmapsto [W \otimes_{o\Delta} o\Gamma]$ (and for $\mathbb{Z}$ by the identity). If $\Delta \subset \Gamma$ the contravariant map is restriction of scalars (and for $\mathbb{Z}$ multiplication by $[\Gamma : \Delta]$). If $F \subset E$ the maps in one direction are given by $[W] \longmapsto [W \otimes_{o\Gamma} o_E\Gamma]$, (with the identity on $\mathbb{Z}$), and in the other direction by restriction of scalars (with multiplication by $[E : F]$ on $\mathbb{Z}$).

(iv) The isomorphisms (cf. Theorem 2)

$$\text{Det}(F\Gamma)^*/\text{Det}(o\Gamma)^* \cong \text{Hom}_{\Omega_F}(R_\Gamma, F_c^*)/\text{Det}(o\Gamma)^* \cong K_oT(o\Gamma), \quad \text{(local case)},$$

$$\left.\begin{array}{l}\mathrm{Det}(JF\Gamma)/\mathrm{Det}(U\mathcal{O}\Gamma) \cong K_oT(\mathcal{O}\Gamma)\ ,\\ \mathrm{Det}(JF\Gamma)/\mathrm{Det}(U\mathcal{O}\Gamma)\cdot\mathrm{Det}(F\Gamma)^* \cong \mathrm{Cl}(\mathcal{O}\Gamma)\ ,\\ \mathrm{Hom}_{\Omega_F}(R_\Gamma,JF_c)/\mathrm{Det}(U\mathcal{O}\Gamma)\cdot\mathrm{Hom}_{\Omega_F}(R_\Gamma,F_c^*) \cong \mathrm{Cl}(\mathcal{O}\Gamma),\end{array}\right\}\text{(global case)},$$

are isomorphisms of Frobenius modules (for Γ and for F). The sequences

$$0 \longrightarrow \mathrm{Cl}(\mathcal{O}\Gamma) \longrightarrow K_o(\mathcal{O}\Gamma) \longrightarrow \mathbb{Z} \longrightarrow 0\ ,$$

$$0 \longrightarrow \mathbb{Z} \longrightarrow K_o(\mathcal{O}\Gamma) \longrightarrow \mathrm{Cl}(\mathcal{O}\Gamma) \longrightarrow 0\ ,$$

are exact sequences of Frobenius modules for the variable Γ and the first sequence also for the variable F.

There are alternative descriptions for some of the functors on modules, associated with change of groups.

Supplement to Theorem 13 (i) Let Δ be a subgroup of Γ. Then the functor on $\mathcal{O}\Delta$-modules $W \longmapsto W \otimes_{\mathcal{O}\Delta} \mathcal{O}\Gamma$ is equivalent to $W \longmapsto \mathrm{Map}_\Delta(\Gamma,W)$

(ii) Let $\Gamma = \Delta/\Phi$, Φ a normal subgroup of Δ. Then on locally free $\mathcal{O}\Delta$-modules and on l.f.p. torsion $\mathcal{O}\Delta$-modules the functor $W \longmapsto W \otimes_{\mathcal{O}\Delta} \mathcal{O}\Gamma$ is equivalent to $W \longmapsto W^\Phi$.

Proof of the Supplement (i) Let $\{\gamma_i\}$ be a right transversal of Δ in Γ. Then for all W the map $f \longmapsto \sum f(\gamma_i^{-1}) \otimes \gamma_i$ is an isomorphism $\mathrm{Map}_\Delta(\Gamma,W) \cong W \otimes_{\mathcal{O}\Delta} \mathcal{O}\Gamma$ of $\mathcal{O}\Gamma$-modules, and is independent of the choice of $\{\gamma_i\}$.

(ii) The trace homomorphism

$$x \longmapsto \sum_{\phi\in\Phi} x\phi\ ,\quad W \to W^\Phi\ ,$$

defined for $\mathcal{O}\Delta$-modules W, yields a homomorphism

$$W \otimes_{\mathcal{O}\Delta} \mathcal{O}\Gamma \to W^{\Phi}$$

of $\mathcal{O}\Gamma$-modules, using the adjoint property of the tensor product. If W is locally free, the latter is an isomorphism. On the other hand, if we have a sequence

$$S:\ 0 \to Y \to X \to M \to 0$$

with Y, X locally free then both $S \otimes_{\mathcal{O}\Delta} \mathcal{O}\Gamma$ and S^{Φ} stay exact, whence $M \otimes_{\mathcal{O}\Delta} \mathcal{O}\Gamma \cong M^{\Phi}$, as required.

Remark: We shall introduce some further Frobenius modules in §6.

Before turning to the proof of the theorem we shall establish

2.1 Lemma: *Let* $T: \Gamma \to GL_n(F)$, $S: \Gamma \to GL_m(F_c)$ *be representations of* Γ, *whence* $S \otimes T: \Gamma \to GL_{mn}(F_c)$ *is also a representation. With* S *extended to an algebra homomorphism* $M_{nq}(F\Gamma) \to M_{mnq}(F_c)$, $S \otimes T$ *extended to an algebra homomorphism* $M_q(F\Gamma) \to M_{mnq}(F_c)$, *we have for* $\sum c_\gamma \gamma \in M_q(F\Gamma)$ *(with* $c_\gamma \in M_q(F)$, *all* γ*) the formula*

$S(\sum(c_\gamma \otimes T(\gamma))\gamma) = (S \otimes T)(\sum c_\gamma \gamma)$.

Proof: Identifying $M_{nq}(F\Gamma) = M_q(F\Gamma) \otimes_F M_n(F)$, we have $S(\sum(c_\gamma \otimes T(\gamma))\gamma) = S(\sum c_\gamma \gamma \otimes T(\gamma)) = \sum c_\gamma\, S(\gamma) \otimes T(\gamma) = (S \otimes T)(\sum c_\gamma \gamma)$.

Proof of Theorem 13. By Lemma 1.2., $\mathrm{Hom}_{\Omega_F}(R_\Gamma, G)$ is a Frobenius module over $R_{\Gamma,F}$, for variable Γ. Next we look at it for varying basefield. Let E and F be fields, E an extension of F. For going up everything is obvious. For going down we must assume that

$[E : F]$ is finite. Let first $\chi \in R_{\Gamma,E}$, $f \in \mathrm{Hom}_{\Omega_F}(R_\Gamma, G)$, $\phi \in R_\Gamma$. Let $\{\sigma\}$ be a right transversal of Ω_E in Ω_F. Then

$$(N_{E/F}(\chi f))(\phi) = \prod_\sigma ((\chi f)(\phi^{\sigma^{-1}}))^\sigma$$

$$= \prod_\sigma f(\bar{\chi}\, \phi^{\sigma^{-1}})^\sigma = \prod_\sigma f(\bar{\chi}^\sigma \phi)$$

$$= \prod_\sigma (\chi^\sigma f)(\phi) = ((t_{E/F}\, \chi)f)(\phi) ,$$

and therefore

$$N_{E/F}(\chi f) = (t_{E/F}\, \chi)f ,$$

which, by (1.13), gives one of the required equations.

Next let $\chi \in R_{\Gamma,F}$, $f \in \mathrm{Hom}_{\Omega_E}(R_\Gamma, G)$, $\phi \in R_\Gamma$ and $\{\sigma\}$ as above. Then

$$(N_{E/F}(\chi f))(\phi) = \prod_\sigma ((\chi f)(\phi^{\sigma^{-1}}))^\sigma$$

$$= \prod_\sigma f(\bar{\chi}\, \phi^{\sigma^{-1}})^\sigma = \prod_\sigma f((\bar{\chi}\phi)^{\sigma^{-1}})^\sigma$$

$$= (N_{E/F}\, f)(\bar{\chi}\phi) = (\chi\, N_{E/F}\, f)(\phi) ,$$

whence

$$N_{E/F}(\chi f) = \chi(N_{E/F}\, f) .$$

We have now seen that $\mathrm{Hom}_{\Omega_F}(R_\Gamma, G)$ is indeed a Frobenius module over $R_{\Gamma,F}$ in the variable F. The remaining assertions under (i) are obvious.

The procedure for proving the results under (ii) is as follows: One has to show for each Det-group that it is an $R_{\Gamma,F}$-submodule of the appropriate Hom-group, for fixed Γ and fixed F. Using the relevant results in the change of ring chapter IV (in particular Theorems 8 and 10), it follows then that we have a sub pre-Frobenius module of a Frobenius module. The assertion then is a consequence of Lemma 1.2, (iii). To take the easiest case, $\mathrm{Det}((F\Gamma)^*)$ is an $R_{\Gamma,F}$-submodule of $\mathrm{Hom}_{\Omega_F}(R_\Gamma, F_c^*)$ by Lemma 2.1, and the fact that $\mathrm{Det}(GL_n(F\Gamma)) = \mathrm{Det}(GL_1(F\Gamma))$. The proof of the corresponding result in the other cases has to be adjourned.

Next we consider the bifunctor $W, Z \longmapsto W \otimes_{\mathcal{O}} Z$, which is the subject of (iii). It is obviously additive in each variable, $W \otimes_{\mathcal{O}} Z$ is exact in W, and we have the associative law $W \otimes_{\mathcal{O}} (Z_1 \otimes_{\mathcal{O}} Z_2) \cong (W \otimes_{\mathcal{O}} Z_1) \otimes_{\mathcal{O}} Z_2$. Thus firstly $\otimes_{\mathcal{O}} Z$ preserves the property of an $\mathcal{O}\Gamma$-module to be free, locally free, or l.f.p.torsion. We also conclude that the Grothendieck groups $K_oT(\mathcal{O}\Gamma)$ and $K_o(\mathcal{O}\Gamma)$ become modules over the ring G, which is the Grothendieck group of $\mathcal{O}\Gamma$-lattices Z with respect to direct sum, made into a ring via the tensor product. The identity of this ring, namely the class of $\mathcal{O}$ with trivial Γ-action, does indeed act as identity. The crucial step is then the proof that the action of Z only depends on the character χ_Z of the $F\Gamma$-module $Z \otimes_{\mathcal{O}} F$. The trivial character $\varepsilon = \chi_{\mathcal{O}}$ is of this form. For $\mathcal{O}$ local in fact all actual characters of $R_{\Gamma,F}$ are of this form, and thus one concludes that $K_o(\mathcal{O}\Gamma)$ and $K_oT(\mathcal{O}\Gamma)$ are $R_{\Gamma,F}$-modules. In the global case the same conclusion is obtained by showing that if χ is the character of an $F\Gamma$-module V then $\chi + \varepsilon = \chi_Z$, for some Z. In fact, V is spanned by some $\mathcal{O}\Gamma$-lattice Z', and there is an ideal Z'' of $\mathcal{O}$ so that $Z = Z' + Z''$ is free over $\mathcal{O}$.

We shall now turn to the proof that the class of $W \otimes_{\mathcal{O}} Z$ only depends on W and on χ_Z. Suppose for the moment that $\mathcal{O}$ is local. Let X be a free $\mathcal{O}\Gamma$-module with basis $\{x_i\}$, $1 \leqslant i \leqslant q$, this being used to identify $\mathrm{End}_{F\Gamma}(X \otimes_{\mathcal{O}} F) = M_q(F\Gamma)$. Let $c \in GL_q(F\Gamma)$ and assume (no loss of generality) that $c \in M_q(\mathcal{O}\Gamma)$. Then

$\mathrm{Det}(c) \mapsto [X/c\,X] \in K_oT(\mathcal{O}\Gamma)$ under the surjection $\mathrm{Hom}_{\Omega_F}(R_\Gamma, F_c^*) = \mathrm{Det}(F\Gamma)^* \to K_oT(\mathcal{O}\Gamma)$. We shall show moreover that $\bar{\chi}_Z\mathrm{Det}(c) \mapsto [X/cX \otimes_{\mathcal{O}} Z]$. This yields the required result for the action of $R_{\Gamma,F}$ on $K_oT(\mathcal{O}\Gamma)$ and shows at the same time that the homomorphism $\mathrm{Det}(F\Gamma)^* \to K_oT(\mathcal{O}\Gamma)$ is one of $R_{\Gamma,F}$-modules. Therefore its kernel $\mathrm{Det}(\mathcal{O}\Gamma)^*$ is an $R_{\Gamma,F}$-module and the local isomorphisms under (iv) preserve the structure of $R_{\Gamma,F}$-modules.

Indeed let $\{u_r\}$ be a free $\mathcal{O}$-basis of Z, used to define the matrix representation T corresponding to χ, by

$$u_r\gamma = \sum_s t_{r,s}(\gamma)\, u_s .$$

Now $\{x_i \otimes u_r\}$ is easily seen to be a free $\mathcal{O}\Gamma$-basis of $X \otimes_{\mathcal{O}} Z$. Write the given matrix c in the form

$$c = \sum_\gamma c_\gamma \gamma , \quad c_\gamma = (c_{k,\ell,\gamma}) \in M_q(\mathcal{O}) .$$

Then we get an $\mathcal{O}\Gamma$-basis $\{y_\ell\}$ of cX, where $y_\ell = \sum_{k,\gamma} x_k\gamma\, c_{k,\ell,\gamma}$ and thus now

$$y_\ell \otimes u_r = \sum_{k,\gamma} (x_k \otimes u_r\, \gamma^{-1})\, \gamma\, c_{k,\ell,\gamma} =$$

$$= \sum_{k,s,\gamma} (x_k \otimes u_s)\, t_{r,s}(\gamma^{-1})\, c_{k,\ell,\gamma}\, \gamma .$$

Thus $(cX \otimes_{\mathcal{O}} Z) = c'(X \otimes_{\mathcal{O}} Z)$, with

$$c' = \sum_\gamma (c_\gamma \otimes \bar{T}(\gamma))\gamma$$

(recall that $\bar{T}(\gamma)$ is the transpose of $T(\gamma^{-1})$). By Lemma 2.1, we have then for any $\phi \in R_\Gamma$,

$$\mathrm{Det}_\phi(c') = \mathrm{Det}_{\bar{\chi}\phi}(c) = (\chi\,\mathrm{Det}_\phi)(c) ,$$

as we had to show.

We now turn to the global case. The action of Z on a torsion module W is compatible with localization. It follows that also globally the ring G acts on $K_0T(o\Gamma)$ via $R_{\Gamma,F}$ and that localization preserves the structure of $R_{\Gamma,F}$-module. Moreover, by taking products or restricted products over local components we deduce that $\mathrm{Det}(JF\Gamma)$ and $\mathrm{Det}(Uo\Gamma)$ are $R_{\Gamma,F}$-submodules of $\mathrm{Hom}_{\Omega_F}(R_\Gamma, JF_c)$, and that the isomorphism for $K_0T(o\Gamma)$ in (iv) preserves $R_{\Gamma,F}$-module structure.

Next one shows that the map $[Y] \mapsto [Y \otimes_o Z]$ on $K_0(o\Gamma)$ only depends on $Z \otimes_o F$. Let Z_1 be another $o\Gamma$-module and free o-module spanning $Z \otimes_o F$. Let X be a free $o\Gamma$-module with $X \otimes_o F = Y \otimes_o F$ and $X \supset Y$. We already know that in $K_0T(o\Gamma)$ we have

$$[X \otimes_o Z / Y \otimes_o Z] = [X \otimes_o Z_1 / Y \otimes_o Z_1] ,$$

whence, in $K_0(o\Gamma)$,

$$[Y \otimes_o Z] - [X \otimes_o Z] = [Y \otimes_o Z_1] - [X \otimes_o Z_1] .$$

But $X \otimes_o Z \cong X \otimes_o Z_1$, both being free $o\Gamma$-modules of the same rank. Therefore finally $[Y \otimes_o Z] = [Y \otimes_o Z_1]$.

Taking next e.g. the identification (in the global case)

$$\mathrm{Cl}(o\Gamma) \cong \mathrm{Cok}(K_0T(o\Gamma) \to K_0(o\Gamma))$$

we obtain the structure of $R_{\Gamma,F}$-module on $Cl(\mathcal{O}\Gamma)$, and we can conclude that all the maps in (iv) preserve the structure of $R_{\Gamma,F}$-module (for fixed Γ and fixed F). For the exact sequences observe that multiplication of a module class [W] by a character χ will multiply its rank by deg χ , and that furthermore multiplication by χ preserves the property of being free over $\mathcal{O}\Gamma$.

By the argument outlined earlier we now deduce that the various Det groups and quotients involving these define Frobenius modules. This applies in particular to the left hand sides in the isomorphisms of (iv) for $K_oT(\mathcal{O}\Gamma)$ and for $Cl(\mathcal{O}\Gamma)$. By the change of order theorems of Chapter IV these isomorphisms are functorial. It follows that $K_oT(\mathcal{O}\Gamma)$ and $Cl(\mathcal{O}\Gamma)$ are Frobenius modules over $R_{\Gamma,F}$.

In an alternative approach one shows directly that the Grothendieck ring $G = G_{\Gamma,F}$ of $\mathcal{O}\Gamma$-lattices defines a Frobenius functor and that $K_o(\mathcal{O}\Gamma)$, $K_oT(\mathcal{O}\Gamma)$ and so $Cl(\mathcal{O}\Gamma)$ are Frobenius modules over it. For group embeddings this is classical and the remaining case of a group surjection is trivial, and so are the corresponding results for change of F . On the other hand $G_{\Gamma,F} \to R_{\Gamma,F}$ is a morphism of Frobenius functors, and in view of the fact that $G_{\Gamma,F}$ acts via $R_{\Gamma,F}$ for each pair Γ,F , the conclusion now follows. The sequence

$$0 \to \mathbb{Z} \to K_o(\mathcal{O}\Gamma) \to Cl(\mathcal{O}\Gamma) \to 0$$

plays an exceptional role - see remark 2 in IV §2. It is not preserved in general under "going down". In the case of restriction to a subgroup Δ the twisting factor in IV Proposition 2.1. however vanishes as $\mathcal{O}\Gamma$ is free over $\mathcal{O}\Delta$. The same is true when the base ring $\mathcal{O}$ to which we go down is a principal ideal ring, e.g. $\mathcal{O}$ local or $\mathcal{O} = \mathbb{Z}$.

§3. Charakter action, Hermitian theory

The conventions are the same as in §1 and §2. The role of

$R_{\Gamma,F}$ is now partly taken over by its subring $R^o_{\Gamma,F} = R_{\Gamma,F} \cap R^o_\Gamma$. Frobenius modules over $R_{\Gamma,F}$ become by restriction Frobenius modules over $R^o_{\Gamma,F}$. We shall also have to introduce certain further Frobenius functors, mainly in connection with a Hermitian analogue to Theorem 13 (iii). As before, we stick again to our three basic cases.

Recall here (cf. [FMc1], [FMc2]) that we can switch from Hermitian forms h over $F\Gamma$ to symmetric bilinear Γ-invariant forms h_1 (with the same underlying module). For given h write

$$h(v,w) = \sum_{\gamma\in\Gamma} h_\gamma(v,w)\gamma^{-1}, \quad h_\gamma(v,w) \in F. \tag{3.1}$$

Then h_1 is indeed Γ-invariant, symmetric bilinear. Conversely given such an h_1 we obtain a Hermitian form h over $F\Gamma$ by setting

$$h(v,w) = \sum_{\gamma\in\Gamma} h_1(v,w\gamma)\gamma^{-1}. \tag{3.2}$$

The two procedures are inverses of each other and are compatible with change of basefield or change of group.

We shall explain this briefly for change of group, working over a field F. h and h_1 are to have the same meaning as above in (3.1), (3.2). Let first Δ be a subgroup of Γ. The given $F\Gamma$-module V is then viewed as a $F\Delta$-module. The Hermitian form over $F\Delta$ is th, with t as in IV (2.13). But trivially

$$(th)_1 = h_1, \tag{3.3}$$

i.e. h_1 is now simply viewed as a Δ-invariant form.

Next let Γ be a subgroup of a group Θ. The new underlying module is $V \otimes_{F\Gamma} F\Theta$. Let $\{\theta_i\}$ be a right transversal of Γ in Θ. For the extension $\tilde{h}$ of h we have

$$\tilde{h}(\sum_i v_i \otimes \theta_i , \sum_i w_j \otimes \theta_j) = \sum_{i,j} \theta_i^{-1} h(v_i,w_j)\theta_j .$$

One quickly computes that

$$\tilde{h}_1(\sum_i v_i \otimes \theta_i , \sum_j w_j \otimes \theta_j) = \sum_i h_1(v_i,w_i) , \tag{3.4}$$

and indeed this is the formula derived directly in [FMc2] .

Finally let now $\Theta = \Gamma/\Phi$, Φ a normal subgroup of Γ . Let Σ be the quotient map $F\Gamma \to F\Theta$. We may identify $V \otimes_{F\Gamma} F\Theta = V^{\Phi}$. Let h' be the restriction of h to V^{Φ} . The Hermitian form $\tilde{h}$ on V^{Φ} over $F\Theta$ is then

$$\tilde{h}(v,w) = \Sigma(h'(v,w)) .$$

Therefore

$$\tilde{h}_1(v,w) = \sum_{\phi\in\Phi} h'_{\phi}(v,w) .$$

(Here $\sum$ is the summation!). In general $h_{\phi}(v,w) = h_1(v,w\phi)$, whence $h'_{\phi} = h'_1$. Thus

$$\tilde{h}_1(v,w) = [\text{order } \Phi]\ h'_1(v,w) . \tag{3.5}$$

Theorem 14. (i) The action of $^-$ induces compatible automorphisms of the Frobenius functor $R_{\Gamma,F}$ and the Frobenius modules $\mathcal{M}$ of Theorem 13. In each case $H^0(\mathcal{M})$ is a Frobenius module over $H^0(R_{\Gamma,F})$, and hence over $R^0_{\Gamma,F}$, where H^0 is defined with respect to the action of the group of order 2, via $^-$.

(ii) The statements in Theorem 13. (i), (ii) remain valid with $R_{\Gamma,F}$, R_{Γ}, Det replaced by $R^{o}_{\Gamma,F}$ R^{s}_{Γ}, Det^{s} respectively. Moreover the restriction maps (cf. II (6.5))

$$\rho_s: \mathrm{Hom}_{\Omega_F}(R_\Gamma, G) \to \mathrm{Hom}_{\Omega_F}(R^s_\Gamma, G)$$

and the maps (cf. II (4.6), (4.7))

$$\mathrm{Nr} : \mathrm{Hom}_{\Omega_F}(R^s_\Gamma, G) \to \mathrm{Hom}_{\Omega_F}(R_\Gamma, G)$$

define homomorphisms of Frobenius bimodules over $R^{o}_{\Gamma,F}$. Thus Ker ρ_s, Cok ρ_s, Ker Nr, Cok Nr are Frobenius bimodules over $R^{o}_{\Gamma,F}$.

(iii) The usual action of characters on Hom-groups (cf. Theorem 13 (i)) gives rise to structures of Frobenius bimodules over $R^{o}_{\Gamma,F}$ on $\mathrm{HCl}(\mathfrak{o}\Gamma)$, $\mathrm{HCl}(F\Gamma)$ and for $\mathfrak{o}$ global on Ad $\mathrm{HCl}(\mathfrak{o}\Gamma)$, Ad $\mathrm{HCl}(F\Gamma)$. The maps $\underline{P}_{\mathfrak{o}\Gamma}$, $\underline{P}_{F\Gamma}$ (global case), the map ν of II. Proposition 6.1, and the maps occurring in II. Proposition 6.2 (ii) and Proposition 6.3 (ii) all define homomorphisms of Frobenius modules.

The proof follows the same lines as that for Theorem 13; in fact things are simpler in that all new actions are defined by the action on Hom-groups. References to Theorems 8 and 10 have to be replaced by references to Theorems 9 and 11, respectively.

The weakness of the last theorem is that it does not cover $K_oH(\mathfrak{o}\Gamma)$ and the discriminant, although it could be extended to $K_oH(F\Gamma)$. There is indeed a ring which maps into $R^{o}_{\Gamma,F}$ and which can be used to define a Frobenius formalism, but it is still an open question whether it is the most general candidate. We shall however describe the action on Hermitian modules and the resulting action on discriminants. We consider representations

$$T: \Gamma \to GL_n(\mathfrak{o}) \tag{3.6}$$

with

$$T(\gamma^{-1}) = T(\gamma)_t \tag{3.7}$$

where, as earlier, the subscript $_t$ denotes transposition. Equivalently we consider an $\mathcal{O}\Gamma$-module Z of the following type. Z has a free basis $\{u_r\}$ over $\mathcal{O}$ and there is a symmetric bilinear form β on $Z \otimes_{\mathcal{O}} F$ which has $\{u_r\}$ as orthonormal basis, i.e., so that

$$\beta(u_r, u_s) = \delta_{r,s} \quad \text{(Kronecker delta)} \tag{3.8}$$

and that furthermore

$$\beta(u\gamma, u'\gamma) = \beta(u,u') \qquad \forall\, \gamma \in \Gamma, \quad \forall\, u,u' \in Z \otimes_{\mathcal{O}} F . \tag{3.9}$$

Let $S(\Gamma,\mathcal{O})$ be the Grohendieck group of such pairs (Z, β), and orthogonal sums. In other words, in $S(\Gamma,\mathcal{O})$ we have

$$(Z_1,\beta_1) + (Z_2,\beta_2) = (Z_1 \oplus Z_2, \beta_1 \perp \beta_2) ,$$

where $(\beta_1 \perp \beta_2)(z_1 + z_2, z_1' + z_2') = \beta_1(z_1,z_1') + \beta_2(z_2,z_2')$, in the obvious notation. For a pair (Z,β) and a Hermitian $\mathcal{O}\Gamma$-module (X,h) , we let Γ act on $X \otimes_{\mathcal{O}} Z$ diagonally, and we define on the $F\Gamma$-module $X \otimes_{\mathcal{O}} Z \otimes_{\mathcal{O}} F$ a form $h * \beta$ into $F\Gamma$ by

$$h * \beta(v \otimes u, v' \otimes u') = \sum_{\gamma \in \Gamma} h_\gamma(v,v') \otimes \beta(u,u'\gamma)\gamma^{-1} \tag{3.10}$$

for $v,v' \in X \otimes_{\mathcal{O}} F$, $u,u' \in Z \otimes_{\mathcal{O}} F$. Here $h_\gamma(v,v')$ is the coefficient in F of γ in the expansion of $h(v,v')$ as in (3.1).

Theorem 15. (i) $S(\Gamma,o)$ is a Frobenius functor in the variable Γ. Associating with each pair (Z,β) the character χ_Z of the representation of Γ on $Z \otimes_o F$, yields a homomorphism

$$S(\Gamma,o) \to R^o_{\Gamma,F}$$

of Frobenius functors. Also extension of base field $F \to F'$ of finite degree, or completion, yields a corresponding homomorphism of rings $S(\Gamma,o) \to S(\Gamma,o')$.

(ii) With (X,h), (Z,β) as above (cf. (3.10)), $(X \otimes_o Z, h * \beta) = (X,h) \cdot (Z,\beta)$ is a Hermitian $o\Gamma$-module. This multiplication defines on $K_0 H(o\Gamma)$ the structure of a contravariant Frobenius module over $S(\Gamma,o)$. Also

$$d((X,h) \cdot (Z,\beta)) = \chi_Z d((X,h)) .$$

Proof: We consider pairs (Y,b), where Y is an $o\Gamma$-lattice and b a non-singular, symmetric, bilinear, Γ-invariant form

$$Y \otimes_o F \times Y \otimes_o F \to F .$$

Let $S'(o,\Gamma)$ be the Grothendieck group of such pairs, up to $o\Gamma$-isomorphism, with respect to orthogonal sum. It becomes a ring under the product

$$(Y,h) \cdot (Y',b') = (Y \otimes_o Y', b \otimes_F b') ,$$

(by abuse of notation we omit a further pair of brackets here) with diagonal action of Γ. If $\Gamma' \to \Gamma$ is a homomorphism of groups, then viewing the (Y,b) over Γ' we get a homomorphism

$$S'(\mathcal{O},\Gamma) \to S'(\mathcal{O},\Gamma') \tag{3.11}$$

of rings.

Next let Γ be a subgroup of a group Θ, and $\{\theta_i\}$ a right transversal of Γ in Θ. Given $(Y,b) \in S'(\mathcal{O},\Gamma)$ we define $(Y \otimes_{\mathcal{O}\Gamma} \mathcal{O}\Theta, \tilde{b}) \in S'(\mathcal{O},\Theta)$ by

$$\tilde{b}(\sum_i v_i \otimes \theta_i, \sum_j w_j \otimes \theta_j) = \sum_i b(v_i,w_i) ,$$

and write $\mathrm{ind}(Y,b) = (Y \otimes_{\mathcal{O}\Gamma} \mathcal{O}\Theta, \tilde{b})$. ind is clearly an additive map $S'(\mathcal{O},\Gamma) \to S'(\mathcal{O},\Theta)$. Denoting by res the homomorphism $S'(\mathcal{O},\Theta) \to S'(\mathcal{O},\Gamma)$ of (3.11), we get (cf. [FMc2]), for $(Y',b') \in S'(\mathcal{O},\Theta)$, the equation

$$(\mathrm{ind}(Y,b)) \cdot (Y',b') = \mathrm{ind}((Y,b) \cdot \mathrm{res}(Y',b')) . \tag{3.12}$$

We need one more result of this type. We let now $\Theta = \Gamma/\Phi$, Φ a normal subgroup of Γ, and consider again a pair $(Y,b) \in S'(\mathcal{O},\Gamma)$. If $V = Y \otimes_{\mathcal{O}} F$, then we may identify, as before, $V^{\Phi} = V \otimes_{F\Gamma} F\Theta$. We now define $\tilde{b}$ on V^{Φ} by

$$\tilde{b}(v,w) = \sum_{\phi\in\Phi} b(v,w\phi) = [\mathrm{order}\ \Phi]b(v,w) . \tag{3.13}$$

Then $(Y^{\Phi},\tilde{b}) \in S'(\mathcal{O},\Theta)$. Moreover, if $(Y',b') \in S'(\mathcal{O},\Theta)$, we clearly may identify

$$((Y \otimes_{\mathcal{O}} Y')^{\Phi}, \widetilde{b \otimes b'}) = (Y^{\Phi} \otimes_{\mathcal{O}} Y', \tilde{b} \otimes b') , \tag{3.14}$$

where on the left (Y',b') is considered "over Γ" (i.e. identifed by abuse of notation with its image under $S'(\mathcal{O},\Theta) \to S'(\mathcal{O},\Gamma)$.

We now observe that the set of particular lattices with orthonormal basis, we considered earlier, is closed under the induction and restriction maps involved in (3.12), and the general contravariant map in (3.11). Therefore indeed it follows that $S(\mathcal{O},\Gamma)$, for varying Γ , is a Frobenius functor.

Next observe that the Hermitian $\mathcal{O}\Gamma$-modules (X,h) correspond biuniquely to pairs (X,h_1) - see (3.1), (3.2). Moreover the product defined in (3.10) satisfies

$$(h*\beta)_1 = h_1 \otimes \beta . \tag{3.15}$$

We use this to prove that $K_0H(\mathcal{O}\Gamma)$ is a contravariant Frobenius module over $S(\mathcal{O},\Gamma)$, using (3.3) and (3.4) to express induction and restriction for pairs (X,h) in terms of the (X,h_1) . The Frobenius formulae for induction-restriction then follow from (3.12). We have to apply this twice. Once with $(Y,b) = (X,h_1)$ and $(Y',b') = (Z,\beta)$, in the obvious notation, and then again with $(Y,b) = (Z,\beta)$, $(Y',b') = (X,h)$. Moreover applying simultaneous restriction to a subgroup, to both (X,h) and (Z,β) (cf. (3.11)) and recalling (3.3),we obtain all the required properties for a group and a subgroup.

To complete the proof of the Frobenius property we have also to consider transition to a quotient group $\Theta = \Gamma/\Phi$. Here we let (X,h) be a Hermitian $\mathcal{O}\Gamma$-module, and $(Z,\beta) \in S(\mathcal{O},\Theta)$ - viewing Z as an $\mathcal{O}\Gamma$-lattice with trivial Φ-action. The required result then follows from (3.5), (3.13) and (3.14), on putting $(Y,b) = (X,h_1)$, and $(Y',b') = (Z,\beta)$.

It only remains to prove the discriminant formula. This we shall do for $\mathcal{O}$ global only.

Let $V = X \otimes_{\mathcal{O}} F$ and let $\{v_i\}$ be a basis of V over $F\Gamma$ and let $b = (h(v_i,v_j))$ be the discriminant matrix. For each p, let a_p be a matrix in $GL_q(F_p\Gamma)$ $(q = \text{rank}_{\mathcal{O}\Gamma}(X))$ which transforms $\{v_i\}$ into a basis $\{x_{i,p}\}$ of X_p over $\mathcal{O}_p\Gamma$. Then $d((X,h))$ is represented in $\text{Det } JF\Gamma \times \text{Hom}_{\Omega_F}(R_\Gamma^s, F_c^*)$ by $(\text{Det}(a), \text{Pf}(b))$, where $a \in JF\Gamma$ has components a_p. Now let $\{u_r\}$ be an orthonormal $\mathcal{O}$-basis of Z. Just as in the proof of Theorem 13 we obtain $a' \in JF\Gamma$, so that for each p, the element a'_p transforms the basis $\{v_i \otimes u_r\}$ into $\{x_{ip} \otimes u_r\}$. Let b' be the discriminant matrix $((h * \beta)(v_i \otimes u_r, v_j \otimes u_s))$. Then $d((X,h)(Z,\beta))$ is represented by $(\text{Det}(a'), \text{Pf}(b'))$. But as for Theorem 13, we get from Lemma 2.1 the equation

$$\text{Det}(a') = \chi\, \text{Det}(a) \qquad (\text{recall here } \chi = \bar{\chi}) . \tag{3.16}$$

On the other hand, using the matrix notation of the proof of Theorem 13, we obtain the equations

$$\begin{aligned}(h * \beta)(v_i \otimes u_r, v_j \otimes u_s) &= \sum_\gamma h_\gamma(v_i,v_j)\, \beta(u_r, u_s\gamma)\gamma^{-1} \\ &= \sum_{\gamma,k} h_\gamma(v_i,v_j)\, \beta(u_r,u_k)\, t_{s,k}(\gamma)\, \gamma^{-1} \\ &= \sum_\gamma h_\gamma(v_i,v_j)\, t_{s,r}(\gamma)\gamma^{-1} , \\ &\text{as } \beta(u_r,u_k) = \delta_{r,k} .\end{aligned}$$

By Lemma 2.1, again

$$\text{Pf}(b') = \chi \text{Pf}(b) .$$

In conjunction with (3.16) this yields the required result.

We denote by $R^{on}_{\Gamma,o}$ ("on" for orthonormal) the image of $S(\Gamma,o)$ in R_Γ (i.e. in $R^{o}_{\Gamma,F}$) . This is a sub Frobenius functor of $R^{o}_{\Gamma,F}$, with respect to the variable Γ , and it has to play a crucial role, if one should want to apply induction methods to $K_0H(o\Gamma)$. Its properties, also in relation to other rings, were studied in [Ri] . In particular Ritter proves

3.1 Proposition: Let F be a number field containing the 4-th roots and the g-th roots of unity (g = order Γ), and let o be its ring of integers. Then actually

$$R^{on}_{\Gamma,o} = R^{o}_{\Gamma} \quad .$$

The condition on F can in general not be seriously weakened. Thus Ritter gives an example of a group Γ , with a splitting field F of $F\Gamma$, such that $R^{on}_{\Gamma,o}$ is of infinite index in $R^{o}_{\Gamma,F}$.

On the other hand, for small fields Ritter obtains a complete description, which roughly says that $R^{on}_{\Gamma,o}$ contains only the obvious characters.

3.2 Proposition (cf. [Ri]). Suppose that either $F = \mathbb{Q}$, or F is a real quadratic field (over $\mathbb{Q}$) . Then the characters in $R^{on}_{\Gamma,o}$ are precisely those of form

$$\chi = \sum_i n_i \operatorname{ind}^{\Gamma}_{\Gamma_i} (\phi_i) \ ,$$

where $n_i \in \mathbb{Z}$, and ϕ_i is a homomorphism $\Gamma_i \to \pm 1$.

Corollary: $R^{on}_{\Gamma,\mathbb{Z}} \supset gR_{\Gamma,\mathbb{Q}}$, for g = order Γ .

§4. Special formulae

The notation is that of the preceding sections. We collect here a number of formulae for determinants and Pfaffians, associated with group characters. These are consequences of our general theory, with $K_{F\Gamma,F}$ now replaced by R_Γ, and they have in fact occurred explicitly or implicitly in the previous work. In the case of number fields there are corresponding formulae with $(F\Gamma)^*$ replaced by $J(F\Gamma)$ not to be stated separately. Throughout Det_χ and Pf_χ are viewed as associated with characters χ and so from our new point of view the dependence on any particular subfield of F_c disappears. We shall use the notations (1.8) - (1.10).

Theorem 16. (i) Let Γ be a quotient group of Δ. If $\mathrm{inf}: R_\Gamma \to R_\Delta$ is the inflation map, $\Sigma: F\Delta \to F\Gamma$ the group ring map induced by $\Delta \to \Gamma$, then

$$\mathrm{Det}_{\mathrm{inf}\chi}(c) = \mathrm{Det}_\chi(\Sigma c), \quad c \in GL_n(F\Delta), \quad \chi \in R_\Gamma,$$

$$\mathrm{Pf}_{\mathrm{inf}\chi}(c) = \mathrm{Pf}_\chi(\Sigma c), \quad c \in H^0 GL_n(F\Delta), \quad \chi \in R_\Gamma^s.$$

(ii) Let Δ be a subgroup of Γ, $\mathrm{res}: R_\Gamma \to R_\Delta$ the restriction map. Then

$$\text{(a)} \begin{cases} \mathrm{Det}_{\mathrm{res}\chi}(c) = \mathrm{Det}_\chi(c), \quad c \in GL_n(F\Delta) \subset GL_n(F\Gamma), \quad \chi \in R_\Gamma, \\ \mathrm{Pf}_{\mathrm{res}\chi}(c) = \mathrm{Pf}_\chi(c), \quad c \in H^0(GL_n(F\Delta)) \subset H^0(GL_n(F\Gamma)), \quad \chi \in R_\Gamma^s. \end{cases}$$

Let $\{\gamma_i\}$ be a left transversal of Δ in Γ, ℓ the F-linear map

$F\Gamma \to F\Delta$, *with* $\ell(\delta) = \delta$ *if* $\delta \in \Delta$ *and* $\ell(\gamma) = 0$ *if* $\gamma \in \Gamma, \gamma \notin \Delta$. *Let* ind: $R_\Delta \to R_\Gamma$ *be the induction map. Then*

$$\text{(b)} \begin{cases} \mathrm{Det}_{\mathrm{ind}\chi}(c) = \mathrm{Det}_\chi(\ell(\gamma_i^{-1} c \gamma_j)), \quad c \in GL_n(F\Gamma), \ \chi \in R_\Delta, \\ \\ \mathrm{Pf}_{\mathrm{ind}\chi}(c) = \mathrm{Pf}_\chi(\ell(\gamma_i^{-1} c \gamma_j)), \quad c \in H^0(GL_n(F\Gamma)), \ \chi \in R_\Delta^s . \end{cases}$$

Here $(\ell(\gamma_i^{-1} c\gamma_j))$ *is the block matrix with* i,j-*block* $\ell(\gamma_i^{-1}\, c\gamma_j)$.

(iii) *Let* L *be a subfield of* F, F *finite separable over* L. *Let* $\{a_r\}$ *be a basis of* F *over* L *with discriminant* $\det t(a_r\, a_s)$, $t = t_{F/L}$ *the trace* $F \to L$. *Then*

$$(N_{F/L}\mathrm{Det}(c))(\chi) = \mathrm{Det}_\chi((t(a_r\, a_s\, c_{ij}))_{(r,i)(s,j)}) \det(t(a_r\, a_s))^{-n\deg(\chi)}$$

$$\text{for } \chi \in R_\Gamma, \ c = (c_{ij}) \in GL_n(F\Gamma),$$

and

$$(N_{F/L}\mathrm{Pf}(c))(\chi) = \mathrm{Pf}_\chi((t(a_r\, a_s\, c_{ij}))_{(r,i)(s,j)}) \det(t(a_r\, a_s))^{-n\deg(\chi)/2},$$

$$\text{for } \chi \in R_\Gamma^s, \ c = (c_{ij}) \in H^0(GL_n(F\Gamma)).$$

Here

$$(N_{F/L}\,\mathrm{Det}(c))\ (\chi) = \prod_{\sigma} (\mathrm{Det}_{\chi^{\sigma^{-1}}}(c))^{\sigma} ,$$

$$(N_{F/L}\,\mathrm{Pf}(c))\ (\chi) = \prod_{\sigma} (\mathrm{Pf}_{\chi^{\sigma^{-1}}}(c))^{\sigma} ,$$

with $\{\sigma\}$ a right transversal of Ω_F in Ω_L .

(iv) $\mathrm{Det}_{\chi\psi}(c) = \mathrm{Det}_{\psi}\ (\sum c_{\gamma}\ T(\gamma)\gamma)$,

for $\chi \in R_{\Gamma,F}$, $\psi \in R_{\Gamma}$, $c = \sum_{\gamma} c_{\gamma}\ \gamma \in GL_q(F\Gamma)$, $c_{\gamma} \in M_q(F)$,

with $T\colon \Gamma \to GL_n\ (F)$ a representation with character χ . Also

$$\mathrm{Pf}_{\chi\psi}(c) = \mathrm{Pf}_{\psi}(\sum c_{\gamma}\ T(\gamma)\gamma) ,$$

with the further conditions on T , ψ , c - as above - that

$$T(\gamma) = T(\gamma)_t^{-1} ,\quad c \in H^{o}(GL_q(F\Gamma)) ,\quad \psi \in R_{\Gamma}^{s} .$$

We mention some applications.

Corollary 1. Let $\gamma \in H^{o}(\Gamma)$, i.e., $\gamma \in \Gamma$, $\gamma^2 = 1$. Let T be a symplectic representation with character $\chi \in R_{\Gamma}^{s}$. The number of eigenvalues -1 of $T(\gamma)$ is even, say $= 2n_{\chi}$ and $\mathrm{Pf}_{\chi}(\gamma) = (-1)^{n_{\chi}}$.

Proof: Take $\Delta = \langle \gamma \rangle$. Then

$$\operatorname{res} \chi = 2n_\chi \phi_\gamma + m_\chi \varepsilon_\gamma$$

where ϕ_γ, ε_γ are the characters of Δ with $\phi_\gamma(\gamma) = -1$, $\varepsilon_\gamma(\gamma) = 1$, and of course ϕ_γ is missing if $\gamma = 1$. By II Proposition 4.6, $\mathrm{Pf}_{2\phi_\gamma}(\gamma) = \mathrm{Det}_{\phi_\gamma}(\gamma)$, hence by (ii) (a) in the Theorem,

$$\mathrm{Pf}_\chi(\gamma) = \mathrm{Pf}_{2\phi_\gamma}(\gamma)^{n_\chi} = \mathrm{Det}_{\phi_\gamma}(\gamma)^{n_\chi} = (-1)^{n_\chi} .$$

Next by (ii) (b) we have, on specialising to elements γ of Γ,

<u>Corollary 2</u>. $\mathrm{Det}_{\mathrm{ind}\chi}(\gamma) = \mathrm{Det}_\chi(\delta_{P(j),i} \cdot \gamma_j^*)_{(i,j)}$, <u>and for</u> $\gamma^2 = 1$ <u>and</u> $\chi \in R_\Delta^s$ <u>also</u> $\mathrm{Pf}_{\mathrm{ind}\chi}(\gamma) = \mathrm{Pf}_\chi(\delta_{P(j),i} \cdot \gamma_j^*)_{(i,j)}$.

<u>Here</u> δ <u>is the Kronecker symbol</u>, P <u>is the permutation induced by</u> γ <u>on the left cosets of</u> Δ <u>in</u> Γ <u>and</u> $\gamma\gamma_j = \gamma_{P(j)}\gamma_j^*$.

<u>Note</u>: The first formula is easily seen to be the known transfer formula for the values of $\mathrm{Det}_{\mathrm{ind}\chi}$ on Γ. Thus (ii) (b) is a generalization of this.

<u>Proof of Theorem 16</u>. Throughout we are using Proposition 1.1 to translate earlier results into character language.

(i) and (iia) follow from Theorem 8 for Det and Theorem 9 for Pf.

To establish (iib) we use Theorem 10 for Det and Theorem 11 for Pf, as well as formulae (2.14) in Theorem 11 for both Det and Pf. The crucial point here is that in the case of induction the matrix $\mathcal{B}$ turning up is actually the identity matrix in our present case.

Next (iii) is again a special case of Theorems 10 and 11, together with their supplements, the latter giving the results in terms of the "norm" $\mathcal{N}_{F/L}$ (taking also account of the change in notation). In addition one has to note that the a_i lie in F and that therefore $\mathrm{Det}_\chi(t\,(a_i\, a_j)) = \mathrm{Det}(t(a_i\, a_j))^{\deg(\chi)}$ and analogously for Pf.

Finally (iv) is a consequence of Lemma 2.1.

§5. Special properties of the group ring involution

Here the underlying field is the field $\mathbb{R}$ of real numbers.

5.1 Proposition: (cf. [We]). The indecomposable involution algebras $(A, \bar{\ })$ which appear as components of $\mathbb{R}\Gamma$ (with $\bar{\gamma} = \gamma^{-1}$ for $\gamma \in \Gamma$) are of the following types, each of which can occur for some Γ.

(i) $A = M_n(\mathbb{R})$, some n, with an orthogonal involution.

(ii) $A = M_n(\mathbb{H})$, some n, $\mathbb{H}$ the real quaternion division algebra, with a symplectic involution.

(iii) $A = M_n(\mathbb{C})$, some n, with some unitary involution, inducing complex conjugation on $\mathbb{C}$.

Proof: The involution on $\mathbb{R}\Gamma$ is positive i.e., $tr(a\,\bar{a}) > 0$ if $a \in \mathbb{R}\Gamma$, $a \neq 0$, where $tr: \mathbb{R}\Gamma \to \mathbb{R}$ is the trace. This property goes over to the indecomposable involution components. If A splits, $A = B \times B^{op}$ then $\exists\, a \neq 0$, e.g., $a = (1,0)$ with $a\,\bar{a} = 0$. Thus (iii) is the only unitary case which can occur. If $A = M_n(\mathbb{R})$ and the involution is symplectic then there is an idempotent $e \neq 0$, with $e\,\bar{e} = 0$. It suffices to consider the case $n = 2$, when

$$e = \begin{pmatrix} 1 & 0 \\ 0 & 0 \end{pmatrix}.$$ If $A = \mathbb{H}$ with orthogonal involution, then one can produce an element a with $a\,\bar{a} < 0$. This extends to $M_n(\mathbb{H})$.

Type (i) occurs for all groups, type (ii) for every quaternion group, type (iii) for every cyclic group of order > 2, always for some suitable n.

Corollary: The discriminant d is surjective if $A = F\Gamma$, $\mathfrak{A} = \mathcal{O}\Gamma$, where F is a local field or a number field, $\mathcal{O}$ its ring of integers (or $\mathcal{O} = F$).

Indeed by the supplement to Theorem 5 in III §3 and by the above Proposition 5.1, d is surjective for $\mathbb{R}\Gamma$. By Theorem 5 this suffices.

§6. Some Frobenius modules

We shall use the apparatus set up earlier in this Chapter to introduce some further Frobenius modules. These will illustrate the theory. Moreover they turn up in arithmetic applications (cf. [CN], [F7]) and they are useful in analysing class groups of group rings.

As before, let $Y(\mathbb{Q}_{p,c})$ be the group of units of the ring of integers (i.e., the integral closure of $\mathbb{Z}_p$) in $\mathbb{Q}_{p,c}$, let $U(E)$ be the group of unit ideles in a number field E (i.e., those having unit components at all finite prime divisors) and let $U(\mathbb{Q}_c)$ be the direct limit, i.e., union of the $U(E)$. - Next we have to introduce a variant of the usual notion of "total positiveness", which is independent of reference to any particular field. Given a real prime divisor p in a number field F we say that an idele a of a number field E containing F is totally positive above p if, for all prime divisors P of E above p, whether real or not, a_P is real and positive. Clearly this property is preserved on extension of E and thus we can speak of an element of $J(\mathbb{Q}_c)$ being totally positive above p.

With F a number field and Γ a finite group, we consider those maps $f \in \mathrm{Hom}_{\Omega_F}(R_\Gamma, U(F_c))$ for which $f(\chi)$ is totally positive above all real prime divisors p of F whenever $\chi \in R_\Gamma^s$. These form a subgroup to be denoted by $\mathrm{Hom}^+_{\Omega_F}(R_\Gamma, U(F_c))$ - strictly speaking one should attach a reference to F in using the symbol +, as indeed the definition involves F directly, not only via Ω_F. No confusion should arise however from this omission. Note also that it will suffice to require $f(\chi)$ to be totally positive above all real prime divisors of F, for irreducible symplectic characters alone.

For the others this will follow by linearity and from the fact that $f(\phi + \bar{\phi})$ will always be totally positive above a real prime divisor of F . (See the proof of Proposition 6.2).

6.1 Proposition: Let A be a maximal order of $F\Gamma$. If o is local then

$$\mathrm{Det}\, A^{*} = \mathrm{Hom}_{\Omega_F}(R_\Gamma, Y(F_c)) .$$

If o is global then

$$\mathrm{Det}\, UA = \mathrm{Hom}^{+}_{\Omega_F}(R_\Gamma, U(F_c)) .$$

Proof: This follows from III. (5.1), (5.2) together with V. Proposition 5.1. The latter ensures that the indecomposable components of $\mathbb{R}\Gamma$ "belonging" to $\mathbb{H}$, i.e., having sign conditions on the reduced norms are precisely those associated with irreducible symplectic characters.

6.2 Proposition: The groups

$$\mathrm{Hom}_{\Omega_F}(R_\Gamma, Y(F_c)) \qquad \text{(local case)} ,$$

$$\mathrm{Hom}^{+}_{\Omega_F}(R_\Gamma, U(F_c)) \qquad \text{(global case)} ,$$

define Frobenius modules over $R_{\Gamma,F}$, both with respect to Γ and with respect to F , subfunctors of $\mathrm{Hom}_{\Omega_F}(R_\Gamma , F_c^{*})$ and of $\mathrm{Hom}_{\Omega_F}(R_\Gamma, J(F_c))$ respectively.

Let A be a maximal order in $F\Gamma$, containing $o\Gamma$. For o

global define the kernel group (cf. [F7])

$$D(\mathcal{O}\Gamma) = \mathrm{Ker}\ [\mathrm{Cl}(\mathcal{O}\Gamma) \to \mathrm{Cl}(A)] \quad . \tag{6.1}$$

By Proposition 6.1 and Theorem 2,

$$D(\mathcal{O}\Gamma) \cong \frac{\mathrm{Hom}^+_{\Omega_F}(R_\Gamma, U(F_c))}{\mathrm{Hom}^+_{\Omega_F}(R_\Gamma, Y(F_c))\ \mathrm{Det}((\mathcal{O}\Gamma)^*)} , \tag{6.2}$$

where here $Y(F_c)$ is the group of global units in F_c, and the $^+$ has the obvious connotation. Thus

$$\mathrm{Hom}^+_{\Omega_F}(R_\Gamma, Y(F_c)) = \mathrm{Hom}^+_{\Omega_F}(R_\Gamma, U(F_c)) \cap \mathrm{Hom}_{\Omega_F}(R_\Gamma, F_c^*) \tag{6.3}$$

inside the group $\mathrm{Hom}_{\Omega_F}(R_\Gamma, J(F_c))$. We also have

$$\mathrm{Ker}\ [K_0T(\mathcal{O}\Gamma) \to K_0T(A)] = \mathrm{Hom}^+_{\Omega_F}(R_\Gamma, U(F_c))/\mathrm{Det}((\mathcal{O}\Gamma)^*) \quad . \tag{6.4}$$

From 6.2 we then get the

Corollary: $D(\mathcal{O}\Gamma)$ and $\mathrm{Ker}\ [K_0T(\mathcal{O}\Gamma) \to K_0T(A)]$ (A maximal) are Frobenius modules over $R_{\Gamma,F}$, both with respect to Γ and with respect to F.

For the Frobenius property with respect to Γ see [E.M], [Mt], [U] .

Proof of 6.2. The fact that (for $\mathcal{O}$ local) the action of $R_{\Gamma,F}$ on $\mathrm{Hom}_{\Omega_F}(R_\Gamma, F_c^*)$ preserves $\mathrm{Hom}_{\Omega_F}(R_\Gamma, Y(F_c))$ is obvious, and similarly for $\mathcal{O}$ global, provided that F has no real prime divisors. For, then the $^+$ can be omitted, i.e., the signature condition is missing. On the other hand if F has a real prime divisor, then $R_{\Gamma,F} \subset R_{\Gamma,\mathbb{R}}$ and hence $R_{\Gamma,F} \subset R_\Gamma^o$, i.e., $R_{\Gamma,F} = R_{\Gamma,F}^o$. As multiplication by R_Γ^o preserves R_Γ^s so does multiplication by $R_{\Gamma,F}$. This then ensures that $R_{\Gamma,F}$ leaves $\mathrm{Hom}^+_{\Omega_F}(R_\Gamma, U(F_c))$ stable.

The Frobenius property with respect to Γ follows from the fact that any group homomorphism $\Delta \to \Gamma$ yields $R_\Gamma^s \to R_\Delta^s$ and that induction also preserves R^s . Going up to extension fields $(F \subset E)$ clearly gives rise to maps $\mathrm{Hom}^+_{\Omega_F}(R_\Gamma, U(F_c)) \to \mathrm{Hom}^+_{\Omega_E}(R_\Gamma, U(F_c))$ and similarly in the local case. Going down, via $N_{E/F}$ preserves the property of maps to take unit values. Thus what is really left to be shown is that in the global case the signature condition is preserved under $N_{E/F}$, and to do this involves a bit of work.

We fix an absolutely normal number field L containing E as well as all values of the characters of Γ . Let then $\chi \in R_\Gamma^s$, p a real prime divisor of F and $f \in \mathrm{Hom}^+_{\Omega_E}(R_\Gamma, U(\mathbb{Q}_c))$. Then

$N_{E/F}f(\chi) \in U(L)$ and we wish to show that if P is a prime divisor of L above p then $N_{E/F}f(\chi)_P$ is real and positive.

Infinite prime divisors are associated with embeddings in $\mathbb{C}$. Fix such an embedding $r\colon L \to \mathbb{C}$ corresponding to P , and use the same letter to denote the P-component map on JL , viewed as a homomorphism $JL \to \mathbb{C}^*$. Given any embedding $t\colon L \to \mathbb{C}$ we write $t_{|E}$ for its restriction to E . Denote by j complex conjugation. We can then find a right transversal W of Gal(L/E) in Gal(L/F) , which is a disjoint union $W_{\mathbb{R}} \cup W_{\mathbb{C}}$, where for $\omega \in W_{\mathbb{R}}$ the embedding $\omega r_{|E}$ is real, i.e., $\omega rj_{|E} = \omega r_{|E}$, and where the elements of $W_{\mathbb{C}}$ occur in pairs ω,μ with $\omega rj = \mu r$, $\omega_{|E} \neq \mu_{|E}$. We first consider

$$f(\chi^{\omega^{-1}})^{\omega} \cdot f(\chi^{\mu^{-1}})^{\mu}\ , \quad (\omega,\mu) \in W_{\mathbb{C}} \times W_{\mathbb{C}}\ , \tag{6.5}$$

for such a pair ω,μ .

Note that $\mu = \eta\omega$ for $\eta \in G(L/F)$. If χ is a real valued character then χ^t is real valued under any embedding $E \to \mathbb{C}$, i.e., $\chi^{\eta\omega r} = \chi^{\omega rj} = \chi^{\omega r}$, whence $\chi^{\eta^{-1}} = \chi$. Replace χ by $\chi^{\omega^{-1}}$. We conclude that $\chi^{\mu^{-1}} = \chi^{\omega^{-1}}$.

Therefore

$$f(\chi^{\omega^{-1}})^{\omega r} \; f(\chi^{\mu^{-1}})^{\mu r} = f(\chi^{\omega^{-1}})^{\omega r} \; f(\chi^{\omega^{-1}})^{\omega rj}$$

$$= N_{\mathbb{C}/\mathbb{R}} \; f(\chi^{\omega^{-1}})^{\omega r} > 0 \; .$$

In other words

$$(f(\chi^{\omega^{-1}})^{\omega} \cdot f(\chi^{\mu^{-1}})^{\mu})^{r} \text{ is real and positive.} \tag{6.6}$$

Next if $\omega \in W_{\mathbb{R}}$, then the corresponding $\mathcal{P}^{\omega^{-1}}_{|E}$ is a real prime divisor of E , hence by hypothesis

$$f(\chi^{\omega^{-1}})^{\omega r} = f(\chi^{\omega^{-1}})_{\mathcal{P}^{\omega^{-1}}}$$

is real and positive, as $\chi^{\omega^{-1}} \in R^s_\Gamma$. Now the product

$$N_{E/F} f(\chi) = \prod_W f(\chi^{\omega^{-1}})^{\omega}$$

is made up of such factors (with $\omega \in W_{\mathbb{R}}$) and of factors (6.5) for which (6.6) holds. Hence finally $(N_{E/F}\, f(\chi))_p = N_{E/F}\, f(\chi)^r$ is real and positive.

We shall now turn to Hermitian properties. We consider again the maps

$$\rho_s : \begin{cases} \mathrm{Hom}_{\Omega_F}(R_\Gamma, Y(F_c)) \to \mathrm{Hom}_{\Omega_F}(R_\Gamma^s, Y(F_c)), & (o \text{ local}), \\ \\ \mathrm{Hom}^+_{\Omega_F}(R_\Gamma, U(F_c)) \to \mathrm{Hom}^+_{\Omega_F}(R_\Gamma^s, U(F_c)), & (o \text{ global}), \end{cases} \tag{6.7}$$

where the $^+$ on the right hand side has the obvious meaning, i.e., we require f to take totally positive values above all real prime divisors p of F .

6.3 <u>Proposition</u>: <u>The groups in</u> (6.7) <u>are Frobenius modules over</u> $R^o_{\Gamma,F}$, <u>both with respect to</u> Γ <u>and with respect to</u> F , <u>and the maps are homomorphisms of such modules</u>.

The proof is analogous to that of 6.2 and will be omitted. Note also that

$$\mathrm{Im}\ \rho_s = \begin{cases} \mathrm{Det}^s\, A^* & (o \text{ local}), \\ \\ \mathrm{Det}^s\, UA & (o \text{ global}), \end{cases} \tag{6.8}$$

if $A = \bar{A}$ is a maximal order. We get

<u>Corollary</u>: <u>Suppose</u> o <u>is global, and</u> $\rho_s = \rho_{s,\Gamma,F}$ <u>as given in</u>

(6.7). Then $\mathrm{Cok}\ \rho_s$ is a Frobenius module over $R^o_{\Gamma,F}$, both with respect to Γ and with respect to F, and so are

$$\mathrm{Im}\ \rho_s/\mathrm{Det}^s U\mathcal{O}\Gamma \quad \text{and} \quad \mathrm{Hom}^+_{\Omega_F}(R^s_\Gamma, U(F_c))/\mathrm{Det}^s U\mathcal{O}\Gamma ,$$

the former a submodule of the latter. Moreover if A is a maximal order, containing $\mathcal{O}\Gamma$, and $\bar{A} = A$ then

$$\mathrm{Ker}[\mathrm{Ad\ HCl}(\mathcal{O}\Gamma) \to \mathrm{Ad\ HCl}(A)] = \mathrm{Im}\ \rho_s/\mathrm{Det}^s U\mathcal{O}\Gamma .$$

Analogously for $\mathcal{O}$ local.

One can also prove that in the above situation

$$\mathrm{Ker}[\mathrm{HCl}(\mathcal{O}\Gamma) \to \mathrm{HCl}(A)]$$

is a Frobenius module, and that various maps defined in II define homomorphisms of such modules. Once we have reached this stage, the proofs (and the statements) are fairly routine.

Remark: The various groups occuring above (in (6.7), (6.8)) as well as $\mathrm{HCl}(A)$, $\mathrm{Ad\ HCl}(A)$ and the maps involving these are defined quite formally for a maximal order A, without the hypothesis that $\bar{A} = A$ and even without the hypothesis that there is an involution invariant maximal order. Proposition 6.3 and its Corollary remain correct in a formal manner, although it is not clear what interpretation to put on these. The same applies to the remainder of this section.

We now come to groups associated with the Brauer decomposition map. See here [Se1] for details on this topic. Let in the sequel ℓ be a fixed prime number. We shall work either in extensions of $\mathbb{Q}_\ell$ and in this case we shall emphasize this by saying that $\mathcal{O} \supset \mathbb{Z}_\ell$, where o is assumed to be local, or we shall work globally. Let $\mathbb{F}_\ell$ be the field of ℓ elements, $\mathbb{F}_{\ell,c}$ its algebraic closure. We shall write $R_{\Gamma,\ell}$ for the Grothendieck group of finitely generated $\mathbb{F}_{\ell,c}$ Γ-modules, or equivalently of classes of matrix representations of Γ over $\mathbb{F}_{\ell,c}$. Two representations T, S give rise to the same class in $R_{\Gamma,\ell}$, if and only if the irreducible representations which appear in their "triangularisation" coincide (in multiplicity). This will imply - extending representations to group rings - that

$$\mathrm{Det}\, T(\lambda) = \mathrm{Det}\, S(\lambda), \quad \lambda \in \mathbb{F}_{\ell,c}\Gamma . \tag{6.9}$$

With o, F as before, and either $o \supset \mathbb{Z}_\ell$ or o global, and with $R_\Gamma = R_{\Gamma,F_c}$, we may find a finite extension E of F, so that every actual character $\chi \in R_\Gamma$ corresponds to a representation $T: \Gamma \to GL_n(o_E)$, o_E the integers in E. Choose once and for all a homomorphism $t: o_E \to \mathbb{F}_{\ell,c}$. Then $t \circ T$ is a representation $\Gamma \to GL_n(\mathbb{F}_{\ell,c})$ and thus yields an element of $R_{\Gamma,\ell}$, only dependent on the character χ of T. We denote it by $d_\ell \chi$. Then $d_\ell: R_\Gamma \to R_{\Gamma,\ell}$ is a homomorphism. It will still depend on the choice of t above, but its kernel

$$\mathrm{Ker}\, d_\ell = \mathrm{Ker}\, d_{\ell,\Gamma}$$

is independent of choices. It can in fact be characterised directly (see [Se1] Corollary 2 to Theorem 42), namely

$$\mathrm{Ker}\, d_\ell = [\chi \in R_\Gamma \mid \chi(\gamma) = 0 \ \forall\ \gamma \in \Gamma \text{ with } (\mathrm{order}\ (\gamma),\ell) = 1]. \tag{6.10}$$

If L is a finite extension of F, o_L the integral closure of o in L, let L_L be the maximal ideal of o_L if o is local (and $F \supset \mathbb{Q}_\ell$), and let L_L be the product of the maximal ideals of o_L above ℓ if o is global. Write $V_\ell(L) = (o_L/L_L)^*$, and let

$V_\ell(F_c)$ be the direct limit of the $V_\ell(L)$. Then Ω_F acts on $V_\ell(F_c)$. If first $\mathfrak{o}$ is local we define a homomorphism

$$r_\ell\colon \mathrm{Hom}_{\Omega_F}(R_\Gamma, Y(F_c)) \to \mathrm{Hom}_{\Omega_F}(\mathrm{Ker}\ d_\ell, V_\ell(F_c)) \qquad (6.11a)$$

where for given f, $r_\ell f$ is the compositum

$$\mathrm{Ker}\ d_\ell \hookrightarrow R_\Gamma \xrightarrow{f} Y(F_c) \xrightarrow{\text{residues}} V_\ell(F_c)\ .$$

Similarly, letting $U(F_c) \to V_\ell(F_c)$ in the global case be the compositum of the semilocal component map $U(F_c) \to U(F_c)_\ell$ and the residue map we get a homomorphism

$$r_\ell\colon \mathrm{Hom}^+_{\Omega_F}(R_\Gamma, U(F_c)) \to \mathrm{Hom}_{\Omega_F}(\mathrm{Ker}\ d_\ell, V_\ell(F_c))\ . \qquad (6.11b)$$

6.4 Proposition: (i) The maps r_ℓ split.

(ii) $\qquad \mathrm{Det}(\mathfrak{o}\Gamma)^* \subset \mathrm{Ker}\ r_\ell$, $\qquad$ ($\mathfrak{o}$ local) ,

$\qquad \mathrm{Det}\ U\mathfrak{o}\Gamma \subset \mathrm{Ker}\ r_\ell$ $\qquad$ ($\mathfrak{o}$ global) .

Remark: The definition of r_ℓ at least in the global case - and the last proposition go back to [F7].

Proof of 6.4. (i) One knows (cf. [Se1] Chapters 14 to 16) that there is an Ω_F-subgroup $\mathrm{Im}\ e_\ell$ of R_Γ so that $\mathrm{Im}\ e_\ell \cap \mathrm{Ker}\ d_\ell = 0$ (this being due to the fact that the Cartan map c_ℓ is injective), and so that furthermore $R_\Gamma/(\mathrm{Im}\ e_\ell + \mathrm{Ker}\ d_\ell)$ is a finite ℓ-group (as $\mathrm{Cok}\ c_\ell$ is a finite ℓ-group). On the other hand the $V_\ell(E)$ are finite groups of order prime to ℓ, hence the same is true for $\mathrm{Hom}_{\Omega_F}(\mathrm{Ker}\ d_\ell, V_\ell(F_c))$. Assertion (i) now follows immediately.

(ii) We prove the local result, the global one following in the same way. Any element of $\mathrm{Ker}\ d_\ell$ is of form $\chi - \psi$, with χ and ψ actual characters, and $d_\ell\chi = d_\ell\psi$. Let T, S be representations over $\mathcal{O}_E$ (for some E) corresponding to χ, and to ψ, respectively. Let $\tilde{T}, \tilde{S}$ be their reductions under a homomorphism $t: \mathcal{O}_E \to \mathbb{F}_{\ell,c}$. Let $\lambda \in \mathcal{O}_E(\Gamma)^*$ and let $\tilde{\lambda}$ be its image under t extended to $\mathcal{O}_E\Gamma \to \mathbb{F}_{\ell,c}\Gamma$. Then - with the appropriate change in notation - we see from (6.9) that

$$\mathrm{Det}\ \tilde{T}(\tilde{\lambda}) = \mathrm{Det}\ \tilde{S}(\tilde{\lambda}) ,$$

i.e., $$\mathrm{Det}_{d_\ell\chi}(\tilde{\lambda}) = \mathrm{Det}_{d_\ell\psi}(\tilde{\lambda}) ,$$

i.e., $$r_\ell\ \mathrm{Det}(\lambda)\ (d_\ell\chi - d_\ell\psi) = 1 ,$$

as we had to show.

By restricting to R_Γ^s and $\mathrm{Ker}\ d_\ell^s = \mathrm{Ker}\ d_\ell \cap R_\Gamma^s$ we get maps r_ℓ^s - simply replacing in (6.11) R_Γ and $\mathrm{Ker}\ d_\ell$ by the above symplectic subgroups. Clearly 6.4 (ii) goes over. Now we get (see also (6.4) and the formulae in the Corollary to 6.3.)

<u>Corollary 1</u>. <u>The maps r_ℓ and r_ℓ^s give rise to homomorphisms</u>

$$\mathrm{Ker}[K_oT(\mathcal{O}\Gamma) \to K_oT(A)] \to \mathrm{Hom}_{\Omega_F}(\mathrm{Ker}\ d_\ell, V_\ell(F_c))$$

$$\mathrm{Ker}[\mathrm{Ad\ HCl}(\mathcal{O}\Gamma) \to \mathrm{Ad\ HCl}(A)] \to \mathrm{Hom}_{\Omega_F}(\mathrm{Ker}\ d_\ell^s, V_\ell(F_c))$$

$$\mathrm{Hom}_{\Omega_F}^+(R_\Gamma^s, U(F_c))/\mathrm{Det}^s(\mathcal{O}\Gamma)^* \to \mathrm{Hom}_{\Omega_F}(\mathrm{Ker}\ d_\ell^s, V_\ell(F_c))$$

where A *is a maximal order and* o *is assumed to be global. Analogously for* o *local.*

Analogous constructions for classgroups (o global) were already given in [F7]. We shall only mention the simplest one. We put

$$E_\ell(o\Gamma) = \mathrm{Hom}^+_{\Omega_F}(\mathrm{Ker}\ d_\ell,\ V_\ell(F_c))/r_\ell\ \mathrm{Hom}^+_{\Omega_F}(R_\Gamma,\ Y(F_c)) .$$

Then we get

Corollary 2. The map r_ℓ *gives rise to a homomorphism*

$$D(o\Gamma) \to E_\ell(o\Gamma) .$$

Now we have

6.5 *Proposition: The groups*

$$\mathrm{Hom}_{\Omega_F}(\mathrm{Ker}\ d_\ell,\ V_\ell(F_c)) \quad \text{and} \quad E_\ell(o\Gamma)$$

define Frobenius modules over $R_{\Gamma,F}$, *and the groups*

$$\mathrm{Hom}_{\Omega_F}(\mathrm{Ker}\ d^s_\ell, V_\ell(F_c))$$

define a Frobenius module over $R^o_{\Gamma,F}$, *always with respect to* Γ *and with respect to* F. *The maps in Corollaries* 1 *and* 2 *define homomorphisms of Frobenius modules.*

For $E_\ell(\mathcal{O}\Gamma)$ and similar groups a proof was given in [CN].

Proof: We give the proof for $\mathrm{Hom}_{\Omega_F}(\mathrm{Ker}\ d_\ell, V_\ell(F))$ only - everything else follows then easily. As can easily be seen by (6.10), $\mathrm{Ker}\ d_\ell$ in indeed an ideal of R_Γ. It is moreover preserved under the maps coming from change of group. The invariance properties under change of field are even more trivial.

Other groups which define Frobenius modules in the context of classgroups were defined in [T1].

§7. Some subgroups of the adelic Hermitian classgroup

Our aim is to describe certain subgroups of $\mathrm{Ad\ HCl}(\mathcal{O}\Gamma)$ which are important in the application to the Galois module structure of rings of algebraic integers. Here $\mathcal{O}$ is the ring of integers in a number field F. To begin with no restrictions are imposed on F.

Let $\mathbb{Q}_+^*$ be the multiplicative group of positive rationals. We have an inclusion of subgroups

$$\mathrm{Hom}_{\Omega_\mathbb{Q}}(R_\Gamma^s, \mathbb{Q}_+^*)\ \mathrm{Hom}_{\Omega_F}(R_\Gamma^s, U(\mathbb{Q}_c)) \subset \mathrm{Hom}_{\Omega_F}(R_\Gamma^s, J(\mathbb{Q}_c)) .$$

On transition to the quotient $\bmod \mathrm{Det}^s(U(\mathcal{O}\Gamma))$, we derive a homomorphism

$$\mathrm{Hom}_{\Omega_\mathbb{Q}}(R_\Gamma^s, \mathbb{Q}_+^*)\ \mathrm{Hom}_{\Omega_F}(R_\Gamma^s, U(\mathbb{Q}_c)) \to \mathrm{Ad\ HCl}(\mathcal{O}\Gamma) \tag{7.1}$$

7.1 Proposition: The image of (7.1) is isomorphic to the direct product

$$\mathrm{Hom}_{\Omega_{\mathbb{Q}}}(R_\Gamma^s, \mathbb{Q}_+^*) \times [\mathrm{Hom}_{\Omega_F}(R_\Gamma^s, U(\mathbb{Q}_c))/\mathrm{Det}^s(U(\mathfrak{o}\Gamma))] .$$

Indeed, as $U(\mathbb{Q}_c) \cap \mathbb{Q}_+^* = 1$, the two groups whose product appears on the left in (7.1) intersect trivially, while $\mathrm{Det}^s(U(\mathfrak{o}\Gamma)) \subset \mathrm{Hom}_{\Omega_F}(R_\Gamma^s, U(\mathbb{Q}_c))$.

For $F = \mathbb{Q}$, i.e., $\mathfrak{o} = \mathbb{Z}$, the above subgroup of the adelic Hermitian classgroup was first considered in [F8]. A refinement of the results of [F8] led in [CN-T1] to the consideration of further subgroups, which we shall presently define. The construction is based on the next Theorem.

Theorem 17. (cf. [CN-T1]). Let the number field F be at most tamely ramified at the prime 2. Then, in the group $\mathrm{Hom}_{\Omega_F}(R_\Gamma^s, U(\mathbb{Q}_c))$, we get the relation

$$\mathrm{Hom}_{\Omega_{\mathbb{Q}}}(R_\Gamma^s, \pm 1) \cap \mathrm{Det}^s(U(\mathfrak{o}\Gamma)) = 1 .$$

From the theorem and Proposition 7.1 we get immediately the

Corollary 1. With F as in Theorem 17, the group

$$\mathrm{Hom}_{\Omega_{\mathbb{Q}}}(R_\Gamma^s, \mathbb{Q}_+^*) \times \mathrm{Hom}_{\Omega_{\mathbb{Q}}}(R_\Gamma^s, \pm 1) = \mathrm{Hom}_{\Omega_{\mathbb{Q}}}(R_\Gamma^s, \mathbb{Q}^*) \tag{7.2}$$

embeds in $\mathrm{Ad\,HCl}(\mathfrak{o}\Gamma)$.

We denote this embedding by σ_F.

Before coming to the proof of Theorem 17, we note some consequences of the last Corollary. Let $\Sigma_{F/F'}$ be the base field extension map (cf. IV, § 1), where F' is a subfield of F. Then the diagram

$$\begin{array}{ccc} & & \mathrm{Ad\ HCl}(\mathcal{O}_{F'}\Gamma) \\ & \nearrow^{\sigma_{F'}} & \\ \mathrm{Hom}_{\Omega_{\mathbb{Q}}}(R_\Gamma^s,\ \mathbb{Q}^*) & & \Big\downarrow \Sigma_{F/F'} \\ & \searrow_{\sigma_F} & \\ & & \mathrm{Ad\ HCl}(\mathcal{O}_F\Gamma) \end{array} \tag{7.3}$$

commutes. Therefore if $g \in \mathrm{Ad\ HCl}(\mathbb{Z}\Gamma)$ falls into $\sigma_{F'}(\mathrm{Hom}_{\Omega_{\mathbb{Q}}}(R_\Gamma^s, \mathbb{Q}^*))$ under the map $\Sigma_{F'/\mathbb{Q}}$, then the same is true with F' replaced by F. We therefore have

Corollary 2. The set union

$$I_\Gamma = \bigcup_F \Sigma_{F/\mathbb{Q}}^{-1} (\sigma_F(\mathrm{Hom}_{\Omega_{\mathbb{Q}}}(R_\Gamma^s,\ \mathbb{Q}^*))) , \tag{7.4}$$

with F running over all number fields which are at most tamely ramified over $\mathbb{Q}$ at the prime 2, is a subgroup of $\mathrm{Ad\ HCl}(\mathbb{Z}\Gamma)$. Indeed

$$I_\Gamma = (\bigcup_F \mathrm{Ker}\ \Sigma_{F/\mathbb{Q}}) \times \sigma_{\mathbb{Q}}(\mathrm{Hom}_{\Omega_{\mathbb{Q}}}(R_\Gamma^s,\ \mathbb{Q}^*)) . \tag{7.5}$$

Projection onto the second factor on the right hand side of (7.5) gives rise to a surjection

$$\xi\colon I_\Gamma \to \mathrm{Hom}_{\Omega_{\mathbb{Q}}}(R_\Gamma^s,\ \mathbb{Q}^*) . \tag{7.6}$$

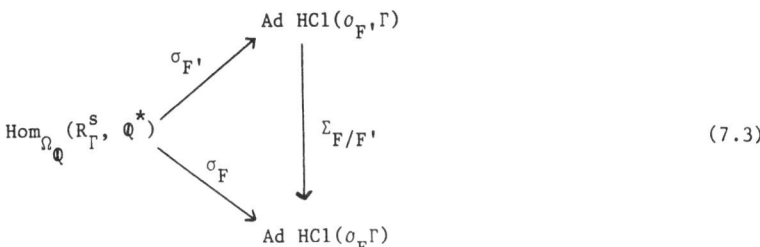

Following this with the projections corresponding to the direct product (7.2), we end up with surjections

$$\left.\begin{aligned} &\eta\colon I_\Gamma \to \mathrm{Hom}_{\Omega_{\mathbb{Q}}}(R_\Gamma^s, \mathbb{Q}_+^*) \;, \\ &\theta\colon I_\Gamma \to \mathrm{Hom}_{\Omega_{\mathbb{Q}}}(R_\Gamma^s, \pm 1) \;. \end{aligned}\right\} \tag{7.7}$$

Next let $(-1)_{\mathrm{fin}}$ be the idele with component -1 at all finite prime divisors and component 1 at infinity. Then we have

<u>Corollary 3.</u> <u>With</u> F <u>as in the theorem</u>

$$\mathrm{Hom}_{\Omega_{\mathbb{Q}}}(R_\Gamma^s, (\pm 1)_{\mathrm{fin}}) \cap \mathrm{Det}^s(U(\mathcal{O}\Gamma)) = 1 \;,$$

<u>where</u> $(\pm 1)_{\mathrm{fin}}$ <u>is the group consisting of</u> 1 <u>and</u> $(-1)_{\mathrm{fin}}$.

<u>Proof</u>: Suppose that the map f lies in the above intersection. Then $f = f_1 f_2$, where $f_1 \in \mathrm{Hom}_{\Omega_{\mathbb{Q}}}(R_\Gamma^s, \pm 1)$, $f_2 \in \mathrm{Hom}_{\Omega_{\mathbb{Q}}}(R_\Gamma^s, (\pm 1)_\infty)$, with $(\pm 1)_\infty$ the group of ideles whose components at infinity are either all -1 , or all $+1$, and whose finite components are 1 . We may enlarge F to a field F' which is totally imaginary, but still tame above $\mathbb{Q}$ at 2 . Then $f_2 \in \mathrm{Det}^s(U(\mathcal{O}'\Gamma))$, $\mathcal{O}'$ the ring of integers in F' . Hence $f_1 \in \mathrm{Det}(U(\mathcal{O}'\Gamma))$, and thus $f_1 = 1$, by the Theorem. Therefore $f = 1$.

<u>Proof of Theorem</u> 17. For simplicity of notation we shall write

$$L_\Gamma = \mathrm{Hom}_{\Omega_{\mathbb{Q}}}(R_\Gamma^s, \pm 1) \cap \mathrm{Det}^s(U(\mathcal{O}\Gamma)) \;,$$

keeping F fixed throughout the proof, but treating Γ as a variable. We then proceed by a series of reduction steps, following [CN-T1] .

1) Reduction to cyclic groups and quaternion groups.

Let Δ be a subquotient of Γ , i.e. a quotient of a subgroup Θ of Γ . By Serre's induction theorem (Theorem 12)

$$R_\Gamma^s = \sum \mathrm{ind}_\Gamma^\Theta \, \mathrm{inf}_\Theta^\Delta \, (R_\Delta^s) \, , \tag{7.8}$$

with Δ running over all cyclic or quaternion subquotients of Γ (quaternion in the generalised sense). We know (cf. Theorem 14) that the dual map $\mathrm{coinf}_\Delta^\Theta \, \mathrm{res}_\Theta^\Gamma$ takes $\mathrm{Hom}_{\Omega_K}(R_\Gamma^s, G)$, and $\mathrm{Det}^s(U(\mathcal{O}_K\Gamma))$, into $\mathrm{Hom}_{\Omega_K}(R_\Delta^s, G)$, and $\mathrm{Det}^s(U(\mathcal{O}_K\Delta))$, respectively, and this for any Ω_K-module G , with either $K = \mathbb{Q}$, or $K = F$.

By (7.8) the map

$$\prod_\Delta \mathrm{coinf}_\Delta^\Theta \, \mathrm{res}_\Theta^\Gamma : L_\Gamma \to \prod_\Delta L_\Delta \, ,$$

with the range of Δ being the same as in (7.8), is injective. It will thus suffice to prove that $L_\Gamma = 1$, for Γ cyclic or quaternion.

2) Reduction to cyclic or quaternion 2-groups.

For the moment Γ can still be quite arbitrary. We consider again the map r_ℓ (cf. (6.11) and its restriction r_ℓ^s . By Proposition 6.4.,

$$r_\ell^s(\mathrm{Det}^s(U(\mathcal{O}\Gamma))) = 1 \ . \tag{7.9}$$

On the other hand, if ℓ is odd, then $+1$ and -1 are distinguished modulo any prime above ℓ, and therefore, writing as before $\mathrm{Ker}\ d_\ell^s = \mathrm{Ker}\ d_\ell \cap R_\Gamma^s$, the map r_ℓ^s yields an injection

$$\mathrm{Hom}_{\Omega_{\mathbb{Q}}}(\mathrm{Ker}\ d_\ell^s, \pm 1) \to \mathrm{Hom}_{\Omega_{\mathbb{Q}}}(\mathrm{Ker}\ d_\ell^s, V_\ell(F_c)) \ .$$

Hence we have, by (7.9),

$$\text{If} \quad f \in L_\Gamma \quad \text{and} \quad \chi \in \mathrm{Ker}\ d_\ell^s \ , \quad \text{then} \quad f(\chi) = 1 \ . \tag{7.10}$$

Suppose now that Γ has a quotient group Γ_2, a 2-group with the property that

$$R_\Gamma^s = \inf_\Gamma^{\Gamma_2} R_{\Gamma_2}^s + \sum_{\ell \neq 2} \mathrm{Ker}\ d_\ell^s \ .$$

Then, by (7.10), the map $\mathrm{coinf}_{\Gamma_2}^{\Gamma}$ yields an injection $L_\Gamma \to L_{\Gamma_2}$. Both quaternion groups and cyclic groups satisfy the above conditions, and this reduces us to quaternion and cyclic 2-groups.

We shall moreover at this stage convert the problem into a local one. Localisation at a prime P of F yields an embedding

$$L_\Gamma \to \mathrm{Hom}_{\Omega_F}(R_\Gamma^s, (\pm 1)_P) \cap \mathrm{Det}^s(\mathcal{O}_P\Gamma^*) \ .$$

In the usual manner we can then go over from the global Galois group Ω_F to the local Galois group Ω_{F_P}. We shall moreover take P to lie above 2. We can now reformulate what we want to prove. Let E be a tame extension of $\mathbb{Q}_2$ of finite degree. Let now R_Γ be the ring of virtual characters of Γ in $(\mathbb{Q}_2)_c$. Denote by L'_Γ the intersection of subgroups $\mathrm{Hom}_{\Omega_E}(R^s_\Gamma, \pm 1)$ and $\mathrm{Det}^s(\mathcal{O}_E\Gamma^*)$ of $\mathrm{Hom}_{\Omega_E}(R^s_\Gamma, (\mathbb{Q}_2)^*_c)$.

We have to show

$$L'_\Gamma = 1 \quad \text{if} \quad \Gamma \quad \text{is a quaternion or cyclic 2-group .} \tag{7.11}$$

3) Proof of (7.11) for a cyclic 2-group Γ

Let $f \in L'_\Gamma$, $f(\chi) = \mathrm{Det}_\chi(\lambda)$ with $\lambda \in \mathcal{O}_E\Gamma^*$, for all $\chi \in R^s_\Gamma$. But now R^s_Γ consists of the virtual characters of form $\theta + \bar{\theta}$, for all $\theta \in R_\Gamma$. Also $\mathrm{Det}_{\theta+\bar{\theta}}(\lambda) = \mathrm{Det}_\theta(\lambda\bar{\lambda})$. Thus, for all $\theta \in R_\Gamma$, $\mathrm{Det}_\theta(\lambda\bar{\lambda}) = \pm 1$. By a theorem of C.T.C.Wall (cf. [W6]), this implies that $\mathrm{Det}_\theta(\lambda\bar{\lambda}) = \mathrm{Det}_\theta(\mu\delta)$, where μ is a 2-power root of unity and $\delta \in \Gamma$. As Γ is Abelian the map $x \longmapsto \mathrm{Det}_\theta(x)$ is injective, whence $\lambda\bar{\lambda} = \mu\delta$. As $E/\mathbb{Q}_2$ is tame, $\mu = \pm 1$. Apply the augmentation $\varepsilon\colon \mathcal{O}_E\Gamma \to \mathcal{O}_E$. We get $\varepsilon(\lambda)^2 = \varepsilon(\lambda\bar{\lambda}) = \mu \in \mathcal{O}_E^*$. Again, as E cannot contain the square root of -1, being tame over $\mathbb{Q}_2$, it follows that $\mu = 1$. Thus $\lambda\bar{\lambda} = \delta$. As therefore $\delta = \bar{\delta}(=\delta^{-1})$, we have $\delta^2 = 1$.

Write $\lambda = \sum_{j=o}^{2^n-1} a_j\gamma^j$, where γ is a generator of Γ and 2^n is the order of Γ. If $\delta \neq 1$, then

$$\sum_{j,k} a_j a_k \gamma^{j-k} = \gamma^{2^{n-1}} \quad . \tag{7.12}$$

Moreover, whenever $\gamma^{j-k} = \gamma^{2^{n-1}}$ then also $\gamma^{k-j} = \gamma^{2^{n-1}}$, while of course the pairs (j,k) and (k,j) are distinct. Therefore the coefficient of $\gamma^{2^{n-1}}$ on the left hand side of (7.12) is even, a contradiction. Hence finally $\lambda\bar{\lambda} = 1$, as we had to show.

4) Proof of (7.11) for a quaternion 2-group Γ

Let $\lambda \in o_E\Gamma^*$, with $\mathrm{Det}_\chi(\lambda) = \pm 1$, for all $\chi \in R_\Gamma^s$. As $2R_\Gamma \subset R_\Gamma^s$ we have $\mathrm{Det}_\theta(\lambda)^4 = 1$, for all $\theta \in R_\Gamma$. By Wall's theorem quoted above (cf. [W6]), $\mathrm{Det}_\theta(\lambda) = \mathrm{Det}_\theta(\mu\delta)$, for a root of unity μ, and for $\delta \in \Gamma$. As $E/\mathbb{Q}_2$ is tame, we must have $\mu = \pm 1$. If $\chi \in R_\Gamma^s$ then $\deg(\chi) \equiv 0 \pmod 2$, whence $\mathrm{Det}_\chi(-1) = (-1)^{\deg(\chi)} = 1$; also $\det_\chi = 1$, i.e. $\mathrm{Det}_\chi(\delta) = 1$. Thus $\mathrm{Det}_\chi(\lambda) = 1$, i.e. we have $L'_\Gamma = 1$.

In conclusion we derive an analogue of Theorem 17 for $\mathrm{HCl}(\mathbb{Z}\Gamma)$. Recall the definition of the map $\mathrm{Tr}\colon R_\Gamma \to R_\Gamma^s$ (cf. II (4.6)), given by $\mathrm{Tr}(\chi) = \chi + \bar{\chi}$. Following [CN-T2] we shall define a homomorphism

$$t_\Gamma^*\colon \mathrm{Hom}_{\Omega_\mathbb{Q}}(R_\Gamma^s/\mathrm{Tr}(R_\Gamma), \pm 1) \times \mathrm{Hom}_{\Omega_\mathbb{Q}}(R_\Gamma^s, \mathbb{Q}^*) \to \mathrm{HCl}(\mathbb{Z}\Gamma). \tag{7.13}$$

First we introduce a homomorphism

$$t'_\Gamma\colon \mathrm{Hom}_{\Omega_\mathbb{Q}}(R_\Gamma^s/\mathrm{Tr}(R_\Gamma), \pm 1) \to \mathrm{Hom}_{\Omega_\mathbb{Q}}(R_\Gamma, J(\mathbb{Q}_c)),$$

giving the local components $(t'_\Gamma f)_P(\chi)$ for all irreducible characters χ, and extending by linearity:

$$(t'_\Gamma f_P)(\chi) = \begin{cases} f(\chi \bmod \mathrm{Tr}(R_\Gamma)) & \text{if } \chi \text{ is symplectic and } P \text{ finite,} \\ 1 & \text{otherwise.} \end{cases}$$

Embedding $i\colon \mathrm{Hom}_{\Omega_\mathbb{Q}}(R_\Gamma^s, \mathbb{Q}^*) \subset \mathrm{Hom}_{\Omega_\mathbb{Q}}(R_\Gamma^s, \mathbb{Q}_c^*)$, we get t_Γ^* as the composite

$$\mathrm{Hom}_{\Omega_\mathbb{Q}}(R_\Gamma^s/\mathrm{Tr}(R_\Gamma), \pm 1) \times \mathrm{Hom}_{\Omega_\mathbb{Q}}(R_\Gamma^s, \mathbb{Q}^*) \xrightarrow{t'_\Gamma \times i}$$

$$\to \mathrm{Hom}_{\Omega_\mathbb{Q}}(R_\Gamma, J(\mathbb{Q}_c)) \times \mathrm{Hom}_{\Omega_\mathbb{Q}}(R_\Gamma^s, \mathbb{Q}_c^*) \to \mathrm{HCl}(\mathbb{Z}\Gamma).$$

Now we have

7.2 Proposition: t_Γ^* is injective (cf. [CN-T2]) .

Proof: Suppose $(f,g) \in \operatorname{Ker} t_\Gamma^*$. By II. Proposition 5.3, this means that there exist

$$h \in \operatorname{Hom}_{\Omega_{\mathbb{Q}}}(R_\Gamma, \mathbb{Q}_c^*), \quad u \in \operatorname{Det}(U(\mathbb{Z}\Gamma)) ,$$

so that

$$h^{-1} = (t'f)u^{-1} \tag{7.14}$$

$$h^S = g \text{ , } h^S \text{ the restriction of } h \text{ to } R_\Gamma^S \text{ .} \tag{7.15}$$

It follows from (7.14) that the values of h^S are unit ideles, positive at all real primes. But by (7.15) these values are in $\mathbb{Q}^*$. Thus $h^S = g = 1$. Applying (7.14) once more, we get

$$(t'f)^S = u^S \in \operatorname{Hom}_{\Omega_{\mathbb{Q}}}(R_\Gamma^S, (\pm 1)_{fin}) \cap \operatorname{Det}^S(U(\mathbb{Z}\Gamma)) .$$

Therefore, by Corollary 3 to Theorem 17, $(t'f)^S = 1$. But, in view of the definition of t' , this implies that $f = 1$.

CHAPTER VI. APPLICATIONS IN ARITHMETIC

This chapter deals with a particular Hermitian module, namely that of the ring of integers in a tame normal extension of a global or local field, viewed as a Galois module, together with the Hermitian form coming from the trace. It was this application which originally motivated the making of a general Hermitian theory.

To sketch the background, we start with the original problem of global Galois module structure for a tame extension N/K of number fields with Galois group Γ. The ring $\mathcal{O}_N$ of integers in N is then locally free over $\mathbb{Z}\Gamma$. It was conjectured by the author and proved by M. Taylor that the class $(\mathcal{O}_N)$ in $\mathrm{Cl}(\mathbb{Z}\Gamma)$ is determined by the values of the Artin root number (constant in the functional equation) $W(\chi)$ for symplectic characters χ of the Galois group Γ. The whole theory is described in detail in [F12]. The converse, namely that $(\mathcal{O}_N)$ determines the values of the symplectic $W(\chi)$, is however false, as was shown early on by the author. Additional algebraic structure is required for the solution of this converse problem, and I guessed early on that this should come from the trace form, and that this should suffice also to determine the local, i.e. Langlands root numbers for symplectic characters χ - their product being the global Artin $W(\chi)$. The conjecture in terms of the new concepts of a Hermitian module and its discriminant was stated in [F11], and was finally proved in [CN-T1], [CN-T2]. It is the purpose of this chapter to describe these results, as well as a number of others. Throughout we shall assume the underlying arithmetic theory of Artin L-functions, and Galois Gauss sums, etc. - giving references to the lesser known aspects. Everything that is required is contained in [F12], with proofs or a detailed literatur list. A useful introduction is [Mr].

In the development of the theory of global Galois module structure for tame extensions four main ingredients are discernible. The first is the Hom description of classgroups, such as $CL(\mathbb{Z}\Gamma)$, and the associated group determinant (cf. [F7]). This has also been dealt with in the present volume. The second is the introduction of the generalised resolvent (again in [F7]) for the actual description of (o_N), and the third is the relation between resolvents and Galois Gauss sum (cf. [F7]). The final ingredient consists essentially of the description of Galois Gauss sums as determinants (cf. [T3]), using their deep arithmetic properties, as well as an important fixed point theorem for determinants (cf. [T2]).

In the Hermitian theory similar four stages can be distinguished.For the first stage the appropriate classgroups had not even been defined previously - they are now given again via a Hom description, using the generalised determinant and the generalised Pfaffian. This was briefly indicated in [F8], [F9] and [F10], but it is in the present volume that all the details and full proofs have been given. The second aspect has again to do with the resolvents, which are used for a description of the Pfaffians in our situation - this will be dealt with here in the subsequent § 1. The third and fourth ingredients are essentially the same as in the original global theory. We shall only give an outline of this, referring back for proofs to earlier literature.

§1. Local theory

To begin with, no restrictions are imposed on the fields considered. Let N/K be a Galois extension of finite degree of fields with Galois group Γ, with $F_c \supset N \supset K \supset F$, K/F of finite degree. Then the trace $t_{N/F}$ defines a Γ-invariant form $N \times N \to F$, given by

$$x,y \longmapsto t_{N/F}(xy) ,$$

and in the usual way (see Chapter V §3 (3.1), (3.2)) this gives rise to a Hermitian form

$$\left.\begin{aligned} &h_{N/K,F}\colon\ N \times N \to F(\Gamma)\ , \\ &h_{N/K,F}(x,y) = \sum_{\gamma} t_{N/F}(x,y^{\gamma^{-1}})\ \gamma\ . \end{aligned}\right\} \tag{1.1}$$

Our first aim is to describe the Pfaffians. This will be done in terms of the generalised resolvents. In the sequel R_Γ and R_Γ^s are defined in terms of representations over F_c. For $a \in N$ and $\chi \in R_\Gamma$ we define the resolvent (cf. [F3], [F6], [F7]) by

$$(a|\chi) = \mathrm{Det}_\chi(\sum a^\gamma\ \gamma^{-1})\ . \tag{1.2}$$

Then we have (cf. [F8])

<u>Theorem 18.</u> <u>Let</u> a <u>be a free generator of</u> N <u>over</u> $K\Gamma$. <u>Then, for all</u> $\chi \in R_\Gamma^s$,

$$\mathrm{Pf}_\chi(h_{N/K,K}\ (a,a)) = (a|\chi)\ .$$

<u>Proof:</u> We verify that in $N\Gamma$

$$\overline{\sum_\gamma a^\gamma\gamma^{-1}} \cdot \sum_\gamma a^\gamma\gamma^{-1} = \sum_{\gamma,\sigma} a^\gamma\ a^{\sigma^{-1}\gamma}\ \sigma$$

$$= \sum_\sigma t_{N/K}\ (a.a^\sigma)\sigma^{-1} = h_{N/K,K}\ (a,a)\ .$$

Now apply II Proposition 4.5.

By Theorem 16 (see also Theorem 11), we now obtain

Corollary 1. *If* $\{c_i\}$ *is a basis of* K/F *and* a *is as in the Theorem, then*

$$\mathrm{Pf}_\chi(h_{N/K,F}(ac_i,ac_j)) = \mathcal{N}_{K/F}(a|\chi)\ \det(t_{K/F}(c_i\ c_j))^{\deg(\chi)/2} ,$$

where of course $\deg(\chi)$ *is even and where* $\mathcal{N}_{K/F}(a|\chi) = \Pi\ (a|\chi^{\sigma^{-1}})^{\sigma}$, $\{\sigma\}$ *running over a right transversal of* Ω_K *in* Ω_F .

From now on we consider local fields. Let until further notice $F = \mathbb{Q}_p$, the rational p-adic field for some prime p . Suppose that N/K is tame. Then the valuation ring $\mathcal{O}_N$ of N is locally free over $\mathcal{O}_K\Gamma$ and over $\mathbb{Z}_p\Gamma$. We thus obtain a Hermitian $\mathcal{O}_K\Gamma$-module $(\mathcal{O}_N, h_{N/K,K})$ and a Hermitian $\mathbb{Z}_p\Gamma$-module $(\mathcal{O}_N, h_{N/K,\mathbb{Q}_p})$ (and of course similarly for intermediary fields). Now we get

Corollary 2. *Let* a *be a free generator of* $\mathcal{O}_N$ *over* $\mathcal{O}_K\Gamma$. *Then* $d((\mathcal{O}_N, h_{N/K,K}))$ *is represented by*

$$\chi \longmapsto (a|\chi) \qquad (\chi \in R_\Gamma^s) .$$

For given K there is a unique power d_K of the prime p , so that the discriminant $\det t_{K/\mathbb{Q}_p}(c_i\ c_j)$ of a basis $\{c_i\}$ of $\mathcal{O}_K$ over $\mathbb{Z}_p$ is of form $d_{K,p}\ v_K$, with v_K a unit of $\mathbb{Z}_p$. By Theorem 11 and its supplement (see also Theorem 16) we then have

Corollary 3. *With* a *as in Corollary 2 and* d_K, v_K *as above,* $d((\mathcal{O}_N, h_{N/K,\mathbb{Q}_p}))$ *is represented by*

$$\chi \longmapsto \mathcal{N}_{K/\mathbb{Q}_p}(a|\chi)(d_{K,p}\ v_K)^{\deg(\chi)/2} ,$$

for all $\chi \in R_\Gamma^s$.

We shall also consider the Hermitian $\mathbb{Z}_p\Gamma$-module $(o_K\Gamma, t_{K/\mathbb{Q}_p})$. Here we extend the trace $t_{K/\mathbb{Q}_p} : K \to \mathbb{Q}_p$ to $K\Gamma \to \mathbb{Q}_p\Gamma$, via its action on the coefficients in the group ring, i.e. $t_{K/\mathbb{Q}_p}(\sum_\gamma a_\gamma\gamma) = \sum t_{K/\mathbb{Q}_p}(a_\gamma)\gamma$, for $a_\gamma \in K$. We then also use the symbol $t_{K/\mathbb{Q}_p}$ for the resulting form

$$\left.\begin{aligned} &K\Gamma \times K\Gamma \longrightarrow \mathbb{Q}_p\Gamma , \\ &(b_1, b_2) \longmapsto t_{K/\mathbb{Q}_p}(b_1 b_2) . \end{aligned}\right\} \tag{1.3}$$

The element of $K_oH(\mathbb{Z}_p\Gamma)$

$$(o_N, h_{N/K,\mathbb{Q}_p}) - (o_K\Gamma, t_{K/\mathbb{Q}_p}) \tag{1.4}$$

lies in $\widetilde{K_oH}(\mathbb{Z}_p\Gamma)$ (recall definition II (5.16)), and we have

Theorem 19. *Let* a *be a free generator of* o_N over $o_K\Gamma$. Then

$$d((o_N, h_{N/K,\mathbb{Q}_p}) - (o_K\Gamma, t_{K/\mathbb{Q}_p}))$$

is represented by

$$\chi \longmapsto \mathcal{N}_{K/\mathbb{Q}_p}(a|\chi)$$

for all $\chi \in R_\Gamma^s$.

<u>Proof</u>: The Hermitian $\mathbb{Z}_p\Gamma$-module $(o_K\Gamma, t_{K/\mathbb{Q}_p})$ is the image under "going down" (cf. IV, §2) of the Hermitian $o_K\Gamma$-module $(o_K\Gamma, m)$ (m = multiplication) (see Example (a) in II, §5), whose discriminant is 1, (by II.Proposition 5.5). The element $(o_N, h_{N/K,K}) - (o_K\Gamma, m)$ is the image in $\widetilde{K_oH(o_K\Gamma)}$ of $(o_N, h_{N/K,K})$ under the map c of II. Proposition 5.6, and it maps onto the element (1.4) under "going down". The theorem now follows from Theorem 11 and its supplement.

Our aim is to find local invariants which have global values. Therefore our next step is to "globalise" the local discriminant of the last theorem. We follow what is essentially a symplectic version of the method involved in I. Proposition 2.3. We fix once and for all an embedding

$$\mathbb{Q}_c \;\rightarrow\; \mathbb{Q}_{p,c} \;,$$

and use this to identify the group R_Γ^s of virtual symplectic characters for representations over $\mathbb{Q}_c$ with that for representations over $\mathbb{Q}_{p,c}$. We also extend the embedding of fields above to a homomorphism

$$s\colon \mathbb{Q}_p \otimes_{\mathbb{Q}} \mathbb{Q}_c = (\mathbb{Q}_c)_p \rightarrow \mathbb{Q}_{p,c} \qquad\qquad (1.5)$$

of $\mathbb{Q}_p$-algebras. The Galois group $\Omega_{\mathbb{Q}}$ acts on $(\mathbb{Q}_c)_p$ via $\mathbb{Q}_c$, and the Galois group $\Omega_{\mathbb{Q}_p}$ acts on $\mathbb{Q}_{p,c}$. There is then a homomorphism

$$s\colon \Omega_{\mathbb{Q}_p} \rightarrow \Omega_{\mathbb{Q}} \qquad\qquad (1.6)$$

such that

$$s(x)^\omega = s(x^{s(\omega)}) \text{ for } x \in (\mathbb{Q}_c)_p, \;\; \omega \in \Omega_{\mathbb{Q}_p} \;.$$

(Recall the discussion prior to I. Proposition 2.3).Via the maps s of (1.5), (1.6) we get a homomorphism

$$\mathrm{Hom}_{\Omega_{\mathbb{Q}}}(R_\Gamma^s, (\mathbb{Q}_c)_p^*) \to \mathrm{Hom}_{\Omega_{\mathbb{Q}_p}}(R_\Gamma^s, (\mathbb{Q}_{p,c})^*) .$$

(Restrict to $\Omega_{\mathbb{Q}_p}$ via s and use the covariance of Hom in the second variable). This is in fact an isomorphism. Composing its inverse with the embedding

$$\mathrm{Hom}_{\Omega_{\mathbb{Q}}}(R_\Gamma^s, (\mathbb{Q}_c)_p^*) \to \mathrm{Hom}_{\Omega_{\mathbb{Q}}}(R_\Gamma^s, J(\mathbb{Q}_c)) ,$$

we end up with an embedding

$$s' : \mathrm{Hom}_{\Omega_{\mathbb{Q}_p}}(R_\Gamma^s, (\mathbb{Q}_{p,c})^*) \to \mathrm{Hom}_{\Omega_{\mathbb{Q}}}(R_\Gamma^s, J(\mathbb{Q}_c)) . \tag{1.7}$$

Now s' yields a homomorphism

$$s'' : \mathrm{Hcl}(\mathbb{Z}_p\Gamma) \to \mathrm{AdHCl}(\mathbb{Z}\Gamma) . \tag{1.8}$$

Denote from now on by $d^*(N/K)$ the image of $d((\mathcal{O}_N, h_{N/K,\mathbb{Q}_p}) - (\mathcal{O}_K\Gamma, t_{K/\mathbb{Q}_p}))$ under s''.
Then we have

Corollary to Theorem 19. $d^*(N/K)$ is represented in $\mathrm{Hom}_{\Omega_{\mathbb{Q}}}(R_\Gamma^s, J(\mathbb{Q}_c))$ by the map f, where for all $\chi \in R_\Gamma^s$,

$$f(\chi)_p = \mathcal{N}_{K/\mathbb{Q}_p}(a|\chi) , \text{ a as in Theorem 19,}$$
$$f(\chi)_q = 1 \text{ for all primes } q \neq p .$$

In order to come to the arithmetic aspects proper of the theory we have to introduce same further concepts from algebraic number theory.

Let F be an extension field of $\mathbb{Q}_p$ of finite degree. Denote by μ the group of complex roots of unity. Consider a continuous homomorphism $\theta : F^* \to \mu$ of finite order. There is then a least non negative integer m so that Ker θ contains the units u of the ring $\mathcal{O}_F$ of integers in F, which satisfy the congruence $u \equiv 1 \pmod{P_F^m}$. We define the <u>conductor</u> $\mathfrak{f}(\theta)$ of θ by

$$\mathfrak{f}(\theta) = P_F^m .$$

Here P_F is the maximal ideal of $\mathcal{O}_F$ and $u \equiv 1 \pmod{P_F^0}$ just means that u is a unit.

Next let ψ_F be the homomorphism of the additive group of F into μ given as the compositum

$$F \xrightarrow{\text{trace}} \mathbb{Q}_p \longrightarrow \mathbb{Q}_p/\mathbb{Z}_p \xrightarrow{\sim} \mathbb{Z}[\tfrac{1}{p}]/\mathbb{Z} \xrightarrow{x \to e^{2\pi i x}} \mu .$$

Choose an element $c \in F^*$ with $c\mathcal{O}_F = \mathfrak{f}(\theta)\mathcal{D}_F$, $\mathcal{D}_F$ the absolute different of $\mathcal{O}_F/\mathbb{Z}_p$. The <u>Gauss sum</u> $\tau(\theta)$ of θ is then defined by

$$\tau(\theta) = \sum_u \theta(uc^{-1})\, \psi_F(uc^{-1}) ,$$

where the sum extends over a complete set $\{u\}$ of representatives of units of $\mathcal{O}_F$, modulo $\mathfrak{f}(\theta)$. This is independent of any choices involved.

We shall now extend these definitions to characters of Galois groups. In this context a character of a Galois group Γ of an extension E/F will also be viewed as a character of Ω_F, via the natural surjection $\Omega_F \to \Gamma$, and the same applies to virtual characters. Note also that class field theory gives us a bijection from the Abelian characters of Galois groups, i.e. actual characters of degree one, to the homomorphisms $\theta: F^* \to \mu$, considered above. We shall denote this bijection by $\phi \longmapsto \theta_\phi$.

We shall view <u>conductors</u> $\mathfrak{f}(F,\chi)$ and <u>Galois Gauss sums</u> $\tau(F,\chi)$ as functions of pairs (F,χ) , with χ a character of Ω_F . For the conductors we have

(1.9)

(i) $\mathfrak{f}(F,\chi+\chi') = \mathfrak{f}(F,\chi)\mathfrak{f}(F,\chi')$, or more precisely, for fixed F, $\chi \longmapsto \mathfrak{f}(F,\chi)$ is a homomorphism from the virtual characters into the ideal group of $\mathcal{O}_F$.

(ii) If ϕ is Abelian then $\mathfrak{f}(F,\phi) = \mathfrak{f}(\theta_\phi)$.

(iii) If $F' \supset F$, χ a character of $\Omega_{F'}$, and $\mathrm{ind}\,\chi$ the induced character of Ω_F , then $\mathfrak{f}(F, \mathrm{ind}\chi) = N_{F'/F}\mathfrak{f}(\chi)d(F'/F)^{\deg(\chi)}$, where $N_{F'/F}$ is the norm and $d(F'/F)$ the relative discriminant.

These properties will determine the conductor uniquely. There is actually an explicit formula depending on the fine structure of ramification. We mention here one special case, the one relevant in our context. Let χ be a character of a Galois group Γ of a tame extension N/K , as we had considered earlier. χ is the character associated with a $\mathbb{C}(\Gamma)$-module V . Let $n\chi$ be the character associated with the fixed submodule V^{Γ_0} of the inertia group $\Gamma_0 \subset \Gamma$. Then

$$\mathfrak{f}(K,\chi) = P_K^{\deg(\chi) - \deg(n\chi)} .$$

Next we come to Galois Gauss sums. These are characterised by the following properties.

$$\left.\begin{array}{ll} \text{(i)} & \text{For fixed } F,\ \chi \mapsto \tau(F,\chi) \text{ is a homomorphism from} \\ & \text{the virtual characters into } \mathbb{C}^*. \\ \text{(ii)} & \text{If } \phi \text{ is Abelian then } \tau(F,\phi) = \tau(\theta_\phi)\ . \\ \text{(iii)} & \text{If } F', F, \chi \text{ are as in (1.9) (iii) and if further-} \\ & \text{more } \deg(\chi) = 0 \text{ then } \tau(F, \operatorname{ind}\chi) = \tau(F',\chi)\ . \end{array}\right\} \quad (1.10)$$

Between these objects we have the important relation

$$|\tau(F,\chi)| = N\mathfrak{f}(F,\chi)^{1/2}\ , \qquad (1.11)$$

where $N\mathfrak{f}$ is the absolute norm and the exponent $1/2$ denotes the positive square root. The <u>local root number</u>, or Langlands constant is then defined by

$$W(F,\chi) = \tau(\bar{\chi})/N\mathfrak{f}(\chi)^{1/2}\ . \qquad (1.12)$$

This enters as a local factor into the global Artin root number. For details see e.g. [Mr] or [F12] .

For the subsequent applications we need a theorem giving special properties of the objects we have defined, in the tame symplectic case. We view characters as having global values; thus $\Omega_{\mathbb{Q}}$ acts on these. We shall in the sequel omit the reference to the underlying local field F in our notation, whenever there is no danger of confusion.

<u>Theorem 20.</u> <u>Let</u> χ <u>be a (virtual) symplectic character of a Galois group of a tame extension</u> E/F . <u>Then</u>

(i) $\mathfrak{f}(\chi)$ <u>is the square of an ideal of</u> $\mathcal{O}_F$, <u>and hence</u> $N\mathfrak{f}(\chi)^{1/2} \in \mathbb{Q}^*$.

(ii) $\tau(\chi) \in \mathbb{Q}^*$, $W(\chi) = \pm 1$.

(iii) For all $\omega \in \Omega_{\mathbb{Q}}$,
$\tau(\chi^\omega) = \tau(\chi)$, $W(\chi^\omega) = W(\chi)$, $\mathfrak{f}(\chi^\omega) = \mathfrak{f}(\chi)$.

(See [F7], or [Mr], or [F12] (Theorem 21). The facts that $W(\chi) = \pm 1$, and that $\mathfrak{f}(\chi^\omega) = \mathfrak{f}(\chi)$, have been known, and have only been included for completeness).

The further progress is based on a close connection between the Galois Gauss sums and the norm resolvents $N(a|\chi)$, first established by the author (cf. [F7]) and then in refined form by M. Taylor (cf. [T3]) . This holds for all (tame) characters and was also of fundamental importance in the global theory of Galois module structure. In our situation it tells us that in the Corollary to Theorem 19, $N_{K/\mathbb{Q}_p}(a|\chi)$ may be replaced by $\tau(K,\chi)$. More precisely, from results in [T3] (see also [F12], Theorem 30 and 31) we have

Theorem 21. With N/K and a as in Theorem 19, the map

$$\chi \longmapsto N_{K/\mathbb{Q}_p}(a|\chi)\ \tau(\chi)^{-1} , \quad \chi \in R_\Gamma^s ,$$

with $N_{K/\mathbb{Q}_p}(a|\chi)$ viewed as an idele, lies in $\mathrm{Det}^s(\mathcal{O}_F\Gamma)$, for some number field F which is at most tamely ramified over $\mathbb{Q}$ at the prime 2 .

From this last theorem, from the Corollary to Theorem 19, and using Corollary 2 to Theorem 17 (in Chapter V) and its notations we now get

Theorem 22. (cf. [CN-T1]) $d^*(N/K)$ lies in I_Γ , and

(i) $\xi d^*(N/K) = [\chi \longmapsto \tau(\chi)]$

(ii) $\eta d^*(N/K) = [\chi \longmapsto N\mathfrak{f}(\chi)^{1/2}]$

(iii) $\theta d^*(N/K) = [\chi \longmapsto W(\chi)]$,

N/K _being a tame extension of number fields._

This then is in particular the Hermitian interpretation of the local tame symplectic root number! The equation (ii) actually goes back to [F8](see Theorem 23 below). We shall also later come back to an analogue of the last theorem at the infinite prime.

We now turn to the conductors themselves. In II (5.14) we defined a homomorphism $HCl(\mathcal{O}_K\Gamma) \to \mathrm{Hom}_{\Omega_K}(R_\Gamma^s, I(K_c))$, for a local field. Write

$$\chi \longmapsto g(\chi)$$

for the image of $d(\mathcal{O}_N, h_{N/K,K})$ under this map, assuming again N/K to be tame.

Theorem 23. (cf. [F8])

For $\chi \in R_\Gamma$, $g(\chi+\bar{\chi}) = \mathfrak{f}(\chi)$.

For $\chi \in R_\Gamma^s$, $g(\chi)^2 = \mathfrak{f}(\chi)$.

From this theorem together with Theorem 20 (i) we deduce that $g(\chi)$ is an ideal of $\mathcal{O}_K$ - something which does depend on the special arithmetic situation.

The theorem follows from Corollary 2 to Theorem 18, in conjunction with a theorem in [F7] (see also [F12] Theorem 24) which asserts that, with a a free generator of $\mathcal{O}_N$ over $\mathcal{O}_K\Gamma$ we have the ideal equation

$$((a|\chi)(a|\bar{\chi})) = \delta(\chi) \ , \quad \text{for all} \quad \chi \in R_\Gamma \ . \tag{1.13}$$

Next we come to two l.f.p. torsion modules, which were first considered by S. Chase (cf.[Ch]) in the context of the Galois module structure of rings of integers. Observe that from II. Proposition 6.1 we obtain a commutative diagram

$$\begin{array}{ccc} & K_oH(\mathcal{O}_K\Gamma) & \\ {}^{d}\swarrow & & \searrow^{d'} \\ HCl(\mathcal{O}_K\Gamma) & \xrightarrow[\nu]{} & K_oT(\mathcal{O}_K\Gamma) \ . \end{array}$$

<u>Theorem 24.</u> <u>Suppose that</u> N/K <u>and</u> a <u>satisfy the hypothesis of Theorem 19. Denote by</u> $\mathcal{D}_{N/K}$ <u>the relative different. Then the element</u> $[\mathcal{D}_{N/K}^{-1} / \mathcal{O}_N]$ <u>of</u> $K_oT(\mathcal{O}_K\Gamma)$ <u>is represented by</u>

$$\chi \longmapsto (a|\chi+\bar{\chi}) \ (\text{for all} \quad \chi \in R_\Gamma) \ ,$$

<u>and</u>

$$[\mathcal{D}_{N/K}^{-1} / \mathcal{O}_N] = \nu d((\mathcal{O}_N, h_{N/K,K})) \ .$$

<u>Proof:</u> As

$$\mathcal{D}_{N/K}^{-1} = [x \in N \mid t_{N/K}(x\mathcal{O}_N) \subset \mathcal{O}_K]$$

it follows from the definition of the map d' that

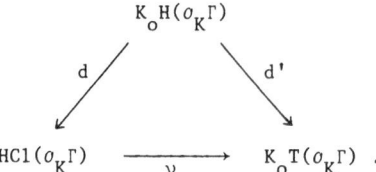

$$[\mathcal{D}^{-1}_{N/K} \,/\, \mathcal{O}_N] = d'((\mathcal{O}_N, h_{N/K,K})) \;.$$

Now use the commutativity of the preceding diagram to get the equation in the theorem. Moreover the map ν is, by its definition, induced from the map $Nr: \chi \longmapsto \chi + \bar{\chi}$. The given representation is thus obtained from the equation already proved and from Corollary 2 of Theorem 18.

The second module was introduced by Chase, as connected with resolvents. We shall first define it, following [Ch] . Then we shall show that it also appears in a Hermitian context.

The N-module $N \otimes_K N$, N acting on the left hand tensor factor has the structure of an $N\Gamma$-module, Γ acting on the right hand tensor factor, i.e. $n(n_1 \otimes n_2) = nn_1 \otimes n_2$, $(n_1 \otimes n_2)^\gamma = n_1 \otimes n_2^\gamma$. As scalar extension of $N = K \otimes_K N$, this module is free over $N\Gamma$ of rank one. Similarly the set $\mathrm{Map}(\Gamma,N)$ of maps $\Gamma \to N$ is an $N\Gamma$-module, where for $f \in \mathrm{Map}(\Gamma,N)$, $n \in N$, $\gamma,\delta \in \Gamma$ we have $(nf)(\delta) = n\cdot f(\delta)$, $f^\gamma(\delta) = f(\gamma\delta)$. Moreover the map

$$n_1 \otimes n_2 \longmapsto [n_1,n_2] \;,\quad [n_1,n_2](\delta) = n_1(n_2^\delta)$$

is an isomorphism

$$N \otimes_K N \cong \mathrm{Map}(\Gamma,N)$$

of $N\Gamma$-modules, which we use to identify the two modules. This $N\Gamma$-module will then contain two $\mathcal{O}_N\Gamma$-lattices, spanning it, namely $\mathrm{Map}(\Gamma,\mathcal{O}_N)$ and $\mathcal{O}_N \otimes_{\mathcal{O}_K} \mathcal{O}_N$, the former containing the latter. $\mathrm{Map}(\Gamma,\mathcal{O}_N)$ is free of rank one over $\mathcal{O}_N\Gamma$, say on a generator e , where $e(1) = 1$, and $e(\gamma) = 0$ for $\gamma \neq 1$. Moreover $\mathcal{O}_N \otimes_{\mathcal{O}_K} \mathcal{O}_N$, as an extension of $\mathcal{O}_N$, is also free of rank one. Thus

$$\mathrm{Map}(\Gamma,\mathcal{O}_N)/\mathcal{O}_N \otimes_{\mathcal{O}_K} \mathcal{O}_N$$

is an l.f.p. torsion module. For the next theorem observe that $\chi \longmapsto (a|\chi)$ does lie in $\mathrm{Hom}_{\Omega_N}(R_\Gamma, K_c^*)$.

Theorem 25. _With_ N/K _and_ a _as in Theorem_ 19,

$$[\mathrm{Map}(\Gamma,\mathcal{O}_N)/\mathcal{O}_N \otimes_{\mathcal{O}_K} \mathcal{O}_N]$$

is represented in $K_oT(\mathcal{O}_N\Gamma)$ _by_

$$\chi \longmapsto (a|\chi) .$$

Also the image of $d((\mathcal{O}_N, h_{N/K,K}))$ _under the ring extension map_ $HCl(\mathcal{O}_K\Gamma) \to HCl(\mathcal{O}_N\Gamma)$ _lies in the image of_ $T\colon K_oT(\mathcal{O}_N\Gamma) \to HCl(\mathcal{O}_N\Gamma)$. _It coincides with_

$$T([\mathrm{Map}(\Gamma,\mathcal{O}_N)/\mathcal{O}_N \otimes_{\mathcal{O}_K} \mathcal{O}_N]) .$$

Corollary: (cf. [F3])

$$d((N, h_{N/K,K})) \in \mathrm{Ker}[HCl(K\Gamma) \to HCl(N\Gamma)] .$$

The Corollary follows from the theorem in conjunction with Theorem 4. It is actually true in a purely field theoretic context.

Proof of Theorem 25. We introduce yet a further free $N\Gamma$-module, namely $N\Gamma$ itself. The map

$$r: f \mapsto \sum_{\gamma} f(\gamma)\gamma^{-1}$$

is an isomorphism $\mathrm{Map}(\Gamma,N) \cong N\Gamma$ of $N\Gamma$-modules. It maps the free generator e of $\mathrm{Map}(\Gamma,\mathcal{O}_N)$, over $\mathcal{O}_K\Gamma$, onto 1, and the free generator $1 \otimes a$ of $\mathcal{O}_N \otimes_{\mathcal{O}_K} \mathcal{O}_N$ onto $\sum a^{\gamma}\gamma^{-1}$. Thus

$$[\mathrm{Map}(\Gamma,\mathcal{O}_N)/\mathcal{O}_N \otimes_{\mathcal{O}_N} \mathcal{O}_N] = [\mathcal{O}_N\Gamma/r(\mathcal{O}_N \otimes_{\mathcal{O}_K} \mathcal{O}_N)]$$

is represented by $\chi \mapsto \mathrm{Det}_{\chi}(\sum a^{\gamma}\gamma^{-1}) = (a|\chi)$, $(\chi \in R_{\Gamma})$. As T is given by the restriction of maps from R_{Γ} to R_{Γ}^{s} (see Theorem 4), the image of the above element under T is indeed represented by $\chi \mapsto (a|\chi)$ $(\chi \in R_{\Gamma}^{s})$ and this yields the full theorem.

From now on we consider an Archimedean local field K, i.e. $K = \mathbb{R}$, the reals, or $K = \mathbb{C}$, the complex numbers. As in (1.8) we get a homomorphism $\mathrm{HCl}(\mathbb{R}\Gamma) \to \mathrm{AdHCl}(\mathbb{Z}\Gamma)$ and we denote again by $d^{*}(N/K)$ the image under this homomorphism of $d((N, h_{N/K,\mathbb{R}}) - (K\Gamma, t_{K/\mathbb{R}}))$. Here of course only three cases can occur, (i) $N = K = \mathbb{C}$, $\Gamma = 1$, (ii) $N = K = \mathbb{R}$, $\Gamma = 1$, (iii) $N = \mathbb{C}$, $K = \mathbb{R}$, Γ of order 2. Only the last case is non-trivial, but our results will hold in all of them.

We will first describe a certain subgroup of $\mathrm{AdHCl}(\mathbb{Z}\Gamma)$. By Corollary 3 to Theorem 17,

$$\mathrm{Hom}_{\Omega_{\mathbb{Q}}}(R_{\Gamma}^{s}, (\pm 1)_{fin}) \cap \mathrm{Det}^{s}(U\mathbb{Z}\Gamma) = 1 .$$

Both these are subgroups of $\mathrm{Hom}_{\Omega_{\mathbb{Q}}}(R_{\Gamma}^{s}, U_{+}(\mathbb{Q}_c))$, where $U_{+}(\mathbb{Q}_c)$ are the unit ideles, real and positive at infinity. This group $U_{+}(\mathbb{Q}_c)$ intersects $\mathbb{Q}^{*}$ trivially. It follows then that the direct product

$$I_{\Gamma}' = \mathrm{Hom}_{\Omega_{\mathbb{Q}}}(R_{\Gamma}^{s}, \mathbb{Q}^{*}) \times \mathrm{Hom}_{\Omega_{\mathbb{Q}}}(R_{\Gamma}^{s}, (\pm 1)_{fin})$$

is in a natural manner a subgroup of $\mathrm{AdHCl}(\mathbb{Z}\Gamma)$. Moreover we have projections

$$\left.\begin{aligned} \eta\colon I'_\Gamma &\longrightarrow \mathrm{Hom}_{\Omega_{\mathbb{Q}}}(R^s_\Gamma, \mathbb{Q}^*) \\ \theta\colon I'_\Gamma &\longrightarrow \mathrm{Hom}_{\Omega_{\mathbb{Q}}}(R^s_\Gamma, (\pm 1)_{fin}) \,. \end{aligned}\right\} \tag{1.14}$$

We now define the root number $W(\chi)$ at infinity, i.e. for all characters χ of $\Gamma = \mathrm{Gal}(N/K)$, K Archimedean. This is additive in χ and has thus to be defined only for irreducible characters. For the identity character ε we set $W(\varepsilon) = 1$. If $\Gamma \neq 1$ then there is precisely one other such character, call it σ . Then $W(\sigma) = -i$. If χ is symplectic then the multiplicity $\langle \chi , \sigma \rangle$ of σ in χ is even, and so

$$W(\chi) = (-1)^{\langle \chi , \sigma \rangle/2} \,. \tag{1.15}$$

Of course, if $\Gamma = 1$ then $W(\chi) = 1$.

Theorem 26. (cf. [F7]

$$d^*(N/K) \in I'_\Gamma \,,$$

and

$$\eta d^*(N/K)(\chi) = W(\chi)$$
$$\theta d^*(N/K)(\chi) = W(\chi)_{fin} \,.$$

Proof: We shall stick to the case $N = \mathbb{C}$, $K = \mathbb{R}$. By Theorem 18, $d^*(N/K)$ is represented by a map f with $f_p = 1$ for all finite p, and

$$f_\infty(\chi) = (a|\chi)$$

where we choose $a = (1-i)/2$ $(i = \sqrt{-1})$. We see immediately that always $(a|\chi) = W(\chi)_\infty$. Thus indeed $d^*(N/K)$ is represented by $\chi \longmapsto W(\chi)_\infty$, and therefore lies in I_Γ'. The rest is now trivial.

§2. The global discriminant

From now on N/K is a tame Galois extension of number fields with Galois group Γ. We recall some facts - for details see e.g. [Mr] or [F12]. If $\chi \in R_\Gamma$, the Artin L-function $\tilde{L}(s,\chi)$ is defined as a product of local factors, one corresponding to each prime divisor p of K, including the infinite ones. The L-function satisfies a functional equation

$$\tilde{L}(s,\chi) = W(\chi)A(\chi)^{1/2-s}\,\tilde{L}(1-s,\bar{\chi}), \tag{2.1}$$

where $A(\chi) > 0$ and $|W(\chi)| = 1$. $W(\chi)$ is the Artin root number, and our interest is focused on this. The first important fact is that it has a decomposition into local factors, established by Langlands, of the form

$$W(\chi) = \prod_{p} W(\chi_p). \tag{2.2}$$

The product running formally over all p, is in fact finite. With N/K we associate a local Galois extension of K_p, whose Galois group Γ_p is embedded in Γ, uniquely to within conjugacy, and

χ_p is the restriction of χ to Γ_p. The $W(\chi_p)$ are then the local root numbers, defined in (1.12) for finite primes, and at the end of §1 also for infinite primes. The global conductor, and the global Galois Gauss sum, respectively, are

$$\left.\begin{aligned} \mathfrak{f}(\chi) &= \prod_p \mathfrak{f}(\chi_p) , \\ \tau(\chi) &= \prod_p \tau(\chi_p) , \end{aligned}\right\} \quad \text{(products over finite primes)} \tag{2.3}$$

and the root number at infinity is

$$W_\infty(\chi) = \prod_p W(\chi_p) \qquad \text{(products over the infinite primes).} \tag{2.4}$$

From (1.12), (2.2) - (2.4) we then have

$$W(\chi) = \tau(\bar{\chi})\, W_\infty(\chi)/N\mathfrak{f}(\chi)^{1/2} . \tag{2.5}$$

We are now turning to the discriminant problem.

<u>Theorem 27.</u> <u>Let</u> N/K <u>be a tame extension of number fields with Galois group</u> Γ. <u>Choose a free generator</u> a <u>of</u> N <u>over</u> $K\Gamma$ <u>and an adele</u> b, <u>so that for all prime divisors</u> p <u>of</u> K, b_p <u>is a free generator of</u> $o_{N,p}$ <u>over</u> $o_{K,p}\Gamma$ <u>(with the usual meaning of these symbols for infinite</u> p).

<u>Then</u> $d((o_N, h_{N/K,K}))$ <u>is represented in</u> $\mathrm{Det}JK\Gamma \times \mathrm{Hom}_{\Omega_K}(R_\Gamma^s, \mathbb{Q}_c^*)$ by (g,h), <u>where</u>

$$g(\chi) = (b|\chi)\cdot(a|\chi)^{-1} , \quad \underline{\text{for all}} \quad \chi \in R_\Gamma ,$$

$$h(\chi) = (a|\chi) , \qquad \underline{\text{for all}} \quad \chi \in R_\Gamma^s .$$

Outline of proof: We use the description of the discriminant (Theorem 3). With the given choice of a and b, let $\lambda \in J_K\Gamma$ be such that $b = a\lambda$. Then from the comparison of the free $\mathcal{O}_K\Gamma$-module $a \cdot \mathcal{O}_K\Gamma$ and of $\mathcal{O}_N$, we conclude that $d((\mathcal{O}_N, h_{N/K,K}))$ is represented by (g,h) where

$$g(\chi) = \mathrm{Det}_\lambda(\chi), \qquad \text{for all } \chi \in R_\Gamma,$$

$$h(\chi) = \mathrm{Pf}_\chi(h_{N/K,K}(a,a)), \text{ for all } \chi \in R_\Gamma^s.$$

By Theorem 18, $h(\chi) = (a|\chi)$. Also (cf. [F7], see [F12] I §4)

$$(a|\chi)\,\mathrm{Det}_\chi(\lambda) = (b|\chi).$$

This is in fact not difficult to prove.

We shall next look at

$$d^*(N/K) = d((\mathcal{O}_N, h_{N/K,\mathbb{Q}}) - (\mathcal{O}_K\Gamma, t_{K/\mathbb{Q}})).$$

From the last theorem and from Theorem 11 and its supplement we have

Corollary: $d^*(N/K)$ *is represented by* (g_1,h_1), *where*

$$g_1(\chi) = \mathcal{N}_{K/\mathbb{Q}}(b|\chi)\cdot\mathcal{N}_{K/\mathbb{Q}}(a|\chi)^{-1}, \quad \textit{for all } \chi \in R_\Gamma,$$

$$h_1(\chi) = \mathcal{N}_{K/\mathbb{Q}}(a|\chi), \qquad \textit{for all } \chi \in R_\Gamma^s.$$

Now recall V, Proposition 7.2, which allows us to view the group

$$I''_\Gamma = \mathrm{Hom}_{\Omega_{\mathbb{Q}}}(R^s_\Gamma/\mathrm{Tr}(R_\Gamma), \pm 1) \times \mathrm{Hom}_{\Omega_{\mathbb{Q}}}(R^s_\Gamma, \mathbb{Q}^*)$$

as a subgroup of $\mathrm{HCl}(\mathbb{Z}\Gamma)$.

Theorem 28. [CN-T2]. *$d^*(N/K)$ lies in I''_Γ . Its component in $\mathrm{Hom}_{\Omega_{\mathbb{Q}}}(R^s_\Gamma/\mathrm{Tr}(R_\Gamma), \pm 1)$ is*

(i) $\chi \longmapsto W(\chi)$,

and its component in $\mathrm{Hom}_{\Omega_{\mathbb{Q}}}(R^s_\Gamma, \mathbb{Q}^*)$ is

$$\chi \longmapsto N\mathfrak{f}(\chi)^{1/2} \cdot W_\infty(\chi) .$$

By (2.5), observing that for $\chi \in R^s_\Gamma$ we have $\chi = \bar{\chi}$ and $W_\infty(\chi) = \pm 1$ (from (2.4) and (1.15)) we get

Corollary 1. *$d^*(N/K)$ is the map $\chi \longmapsto \tau(\chi)$ in I''_Γ .*

In fact the proof of the theorem proceeds via the assertion of the Corollary.

Denote the map (i) of the theorem by $W_{N/K}$, and write $U_{N/K}$ for the class of $\mathcal{O}_N$ in $\mathrm{CL}(\mathbb{Z}\Gamma)$. In V §7, following (7.13), we defined a homomorphism

$$t'_\Gamma\colon \mathrm{Hom}_{\Omega_{\mathbb{Q}}}(R^s_\Gamma/\mathrm{Tr}(R_\Gamma), \pm 1) \to \mathrm{Hom}_{\Omega_{\mathbb{Q}}}(R_\Gamma, J(\mathbb{Q}_c)) .$$

Denote its composition with the surjection

$$\mathrm{Hom}_{\Omega_{\mathbb{Q}}}(R_\Gamma, J(\mathbb{Q}_c)) \to \mathrm{Cl}(\mathbb{Z}\Gamma)$$

(cf. Theorem 2 (VII)) by t_Γ . From Theorem 28 and II. Proposition 6.3 we conclude

Corollary 2. $t_\Gamma W_{N/K} = U_{N/K}$.

This is the main theorem of global Galois module structure, conjectured by the author and proved by Taylor (cf. [T3]). Its proof occupies a mayor portion of [F12], specifically most of its Chapter IV. Theorem 28 does not present a short cut to this result, as the basic ingredients of the proof of the Hermitian theorem really come from [T3].

We shall give a brief outline of the proof strategy for Theorem 28, but will again refer to the appropriate literatur for the required arithmetic results. We shall use the description of $HCl(\mathbb{Z}\Gamma)$ in II. Proposition 5.3. We start with the representative (g_1,h_1) of $d^*(N/K)$, given in the Corollary of Theorem 27. By a theorem in [F7] (see [F12] Theorem 5B or Theorem 20), the map k , defined by $k(\chi) = \mathcal{N}_{K/\mathbb{Q}}(a|\chi)^{-1}\tau(\chi)W'(\chi)$ lies in $\mathrm{Hom}_{\Omega_\mathbb{Q}}(R_\Gamma, \mathbb{Q}_c^*)$. Here $W'(\chi) = W(\chi)$ if χ is irreducible symplectic, $W'(\chi) = 1$ if χ is irreducible non-symplectic, and is extended by linearity. We may thus replace (g_1,h_1) by (g_1k^{-1}, h_1k^s) . We have $(h_1k^s)(\chi) = \tau(\chi)W(\chi) = N\mathfrak{f}(\chi)^{1/2}\, W_\infty(\chi)$, for $\chi \in R_\Gamma^s$. On the other hand $g_1k^{-1}(\chi) = \mathcal{N}_{K/\mathbb{Q}}(b|\chi)\tau(\chi)^{-1}\, W'(\chi)$ lies in the small subgroup $\mathrm{Hom}^+_{\Omega_\mathbb{Q}}(R_\Gamma^s, U(\mathbb{Q}_c))$ (cf. [F7], see [F12] Theorem 5); the aim was then to show that $g_1k^{-1}\cdot t'_\Gamma W_{N/K}$ lies in $\mathrm{Det}U\mathbb{Z}\Gamma$. This was accomplished in [T3] .

In conclusion we mention that Theorems 23, 24 and 25 hold, appropriately reworded, also in the global context. The free generator a of $\mathcal{O}_N$ over $\mathcal{O}_K\Gamma$, appearing in the local theorems has of course now to be replaced by the adelic generator b , which appeared in Theorem 27. The proofs of Theorems 23 and 24 in their global version go by immediate reduction to the local case. This also applies to Theorem 25, except that here an extra step is involved, which we shall now indicate.

We start with the representative (g,h) of $d((\mathcal{O}_N, h_{N/K,K}))$ of Theorem 27. This also represents the lifted discriminant $d((\mathcal{O}_N \otimes_{\mathcal{O}_K} \mathcal{O}_N, N \otimes_K h_{N/K,K}))$. But $\mathrm{Det}(\sum a^\gamma \gamma^{-1}): \chi \longmapsto (a|\chi)$, all $\chi \in R_\Gamma$, lies in $\mathrm{Det} N\Gamma^*$. Therefore

$$(g\cdot\mathrm{Det}(\sum a^\gamma \gamma^{-1}),\ h\cdot\mathrm{Det}^s(\sum a^\gamma \gamma^{-1})^{-1})$$

also represents the lifted discriminant in $\mathrm{HCl}(\mathcal{O}_N\Gamma)$. But $h\cdot\mathrm{Det}^s(\sum a^\gamma \gamma^{-1})^{-1} = 1$, and $g\cdot\mathrm{Det}(\sum a^\gamma \gamma^{-1})$ is the map $\chi \longmapsto (b|\chi)$. Thus the class represented by the above representative does lie in $\mathrm{Im}\,T$. Now we proceed as before.

Literature List

[Ba1] Bass, H.: Lectures on topics in algebraic K-theory, Tata 1967

[Ba2] Bass, H.: Algebraic K-theory, Benjamin, New York 1968

[Bo] Borel, A.: Linear algebraic groups, Benjamin, New York 1969

[Bo-S] Borel, A., Serre, J-P.: Sur certains sous groupes des groupes de Lie compacts, Comm. Math. Helv. 27 (1953), 128 - 139

[Bu] Bushnell, C.J.: A note on involutory division algebras of the second kind, Mathematika 24 (1977), 34 - 36

[Ch] Chase, S.: Ramification invariants and torsion Galois module structure in number fields, to appear in J. of Algebra

[CN] Cassou-Noguès, Ph.: Module de Frobenius et structure Galoisienne des anneaux d'entiers, J. of Algebra 71 (1981), 268 - 289

[CN-T1] Cassou-Noguès, Ph., Taylor, M.J.: Local root numbers and Hermitian Galois module structure of rings of integers, Math. Ann. 263 (1983), 251 - 261

[CN-T2] Cassou-Noguès, Ph., Taylor, M.J.: Constante de l' équation fonctionelle de la fonction L d'Artin d'une représentation symplectique et modérée, Ann. Inst. Fourier Grenoble 33, 2 (1983), 1 - 17

[CR] Curtis, C.W., Reiner, I.: Methods of representation theory, Vol. I, Wiley 1981

[D] Dieudonné, J.: La géometrie des groupes classiques, 2nd ed., Springer, Berlin 1963

[EM] Endo, S., Miyata, T.: Quasi permutation modules over finite groups II, Journ. Math. Soc. Japan 26 (1974), 698 - 713

[F1] Fröhlich, A.: Discriminants of algebraic number fields, Math. Z. 74 (1960), 18 - 28

[F2] Fröhlich, A.: Ideals in an extension field as modules over the algebraic integers in a finite number field, Math. Z. 74 (1960), 29 - 38

[F3] Fröhlich, A.: Resolvents, discriminants and trace invariants, J. of Algebra 4 (1966), 643 - 662

[F4] Fröhlich, A.: Quadratic forms 'a la' local theory. Proc. Camb. Phil. Soc. 63 (1967), 579 - 586

[F5] Fröhlich, A.: Orthogonal and symplectic representations of groups, Proc. London Math. Soc. (3) 24 (1972), 470 - 506

[F6] Fröhlich, A.: Resolvents and Traceform, Math. Proc. Camb. Phil. Soc. 78 (1975), 185 - 210

[F7] Fröhlich, A.: Arithmetic and Galois module structure for tame extension, Crelle 286/287 (1976), 380 - 440

[F8] Fröhlich, A.: Symplectic local constants and Hermitian Galois module structure, in Internat. Symp. Kyoto (ed. S. Iyanaga), Japan Soc. for the Promotion of Science, Tokyo, 1977, 25 - 42

[F9] Fröhlich, A.: Local Hermitian group modules, Quadratic form conference, Queen's papers on Pure and Appl. Math., 46, Kingston, Ontario 1977

[F10] Fröhlich, A.: The Hermitian classgroup, In "integral representations and applications", Proc. Oberwolfach (Ed. K.W. Roggenkamp), Springer Lecture Notes 882 (1981), 191 - 206

[F11] Fröhlich, A.: On parity problems, Sem. théorie des nombres, Univ. Bordeaux (1979)

[F12] Fröhlich, A.: Galois module structure of algebraic integers, Ergebnisse der Mathematik, 3. Folge, Band 1, Springer 1983

[FMc1] Fröhlich, A., McEvett, A.M.: Forms over rings with involution, J. of Algebra 12 (1969) 79 - 104

[FMc2] Fröhlich, A., McEvett, A.M.: The representation of groups by automorphisms of forms, J. of Algebra 12 (1969), 114 - 133

[K] Kneser, M.: Lectures on Galois cohomology of classical groups, Tata 1969

[L] Lam, T.Y.: Induction theorems for Grothendieck groups and Whitehead groups of finite groups, Ann. Ec. Norm. Sup. 1 (1968), 91 - 148

[Mr] Martinet, J.: Character theory and Artin L-functions, in "Algebraic Number Fields" Durham Proceedings, Acad. Press, London 1977, 1 - 87

[Mt] Matchett, A.: Bimodule-Induced homomorphisms of locally free classgroups, J. of Algebra 44 (1977), 196 - 202

[Mc] McEvett, A.M.: Forms over semi-simple algebras with involution, J. of Algebra 12 (1969), 105 - 113

[Ri] Ritter, J.: On orthogonal and orthonormal characters, J. of Algebra 76 (1982), 519 - 531

[Se1] Serre, J-P.: Représentations linéaires des groupes finis, 2e edition, Paris 1971

[Se2] Serre, J-P.: Conducteurs d'Artin des caractères réels, Inv. Math. 14 (1971), 173 - 183

[SE] Swan, R.G., Evans, E.G.: K-theory of finite groups and orders, Lecture Notes in Mathematics 149, Springer, Berlin - New York, 1970

[Sw] Swan, R.G.: Induced representations and projective modules, Ann. of Math. 71 (1960), 552 - 578

[T1] Taylor, M.J.: Locally free classgroups of groups of prime power order, J. of Algebra 50 (1978), 463 - 487

[T2] Taylor, M.J.: A logarithmic approach to classgroups of integral group rings, J. of Algebra 66 (1980), 321 - 353

[T3] Taylor, M.J.: On Fröhlich's conjecture for rings of integers of tame extensions, Inv. Math. 63 (1981), 41 - 79

[U] Ullom, S.V.: Character action on the classgroup of Fröhlich, Preprint 1977

[Wa1] Wall, C.T.C.: On the commutator subgroups of certain unitary groups, J. of Algebra 27 (1973), 306 - 310

[Wa2] Wall, C.T.C.: On the classification of Hermitian forms, II Semisimple Rings, Inv. Math. 18 (1972), 119 - 141

[Wa3] Wall, C.T.C.: On the classification of Hermitian forms, IV Adele Rings, Inv. Math. 23 (1974), 241 - 260

[Wa4] Wall, C.T.C.: On the classification of Hermitian forms, VI Group Rings, Ann. of Math. 103 (1976), 1 - 86

[Wa5] Wall, C.T.C.: On the classification of Hermitian forms, III Complete Semilocal Rings, Inv. Math. 19 (1973), 59 - 71 - V Global Rings, Inv. Math. 23 (1974), 261 - 288

[Wa6] Wall, C.T.C.: Norms of units in group rings, Proc. Am. Math. Soc. 29 (1974), 593 - 632

[We] Weil, A.: Algebras with involution and the classical groups, J. Ind. Math. Soc. 24 (1961), 589 - 623

[Wi] Wilson, S.M.J.: Reduced norms in the K-theory of orders, J. of Algebra 46 (1977), 1 - 11

List of Theorems

Some Further Notation

Index